THE SECRET LIFE OF
THE BRAIN

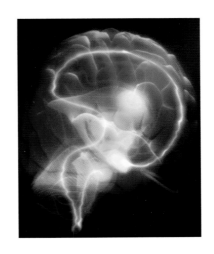

THE SECRET LIFE OF
THE BRAIN

Richard Restak, M.D.

A CO-PUBLICATION OF
THE DANA PRESS AND JOSEPH HENRY PRESS

THE DANA PRESS

JOSEPH HENRY PRESS

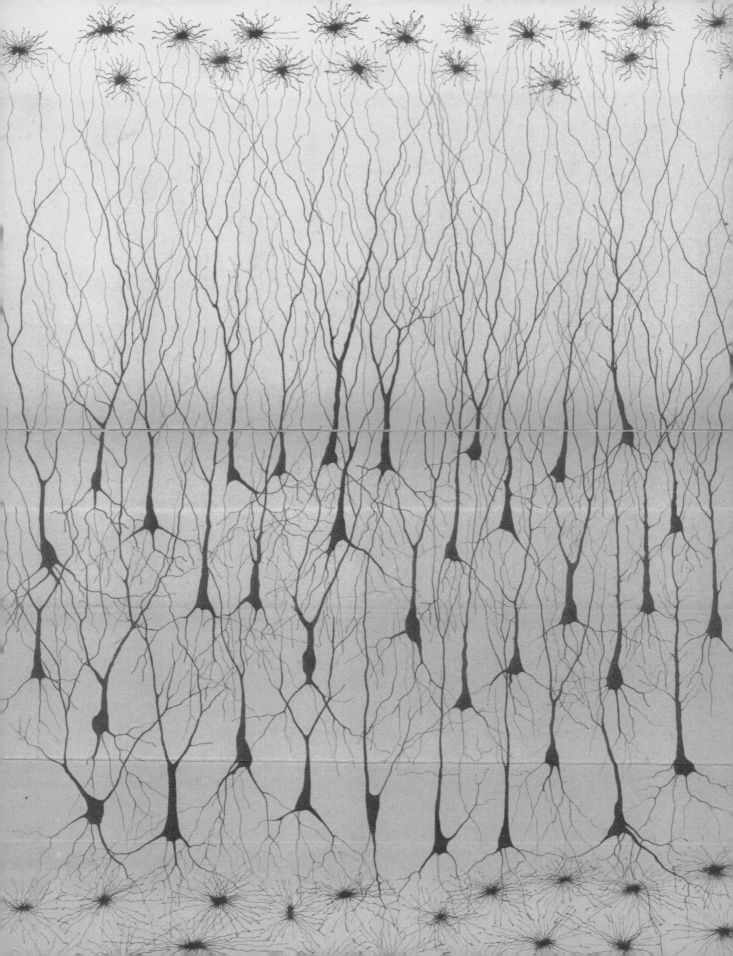

FOREWORD

Science was never my strong suit. As a student, the mathematics and graphs, the test tubes and pipettes, all the wizardry that transforms science from what someone once called "making nasty smells in a laboratory" into a dynamic engine of discovery left me baffled and confused. I was interested in the humanities—in poems and stories which seemed to me to be asking the essential questions about what it is to be human. Perhaps that is why I was drawn to making a series of films on the brain, for no other organ can tell us as much about who we are and what we might become. But now, with the television documentary *The Secret Life of the Brain* finished, when people ask me what I learned about the brain, I begin by telling them what I found out about science.

The first thing I learned was how a scientist confronts a problem. When I asked neuroscientist Susan McConnell, for example, how, out of the union of sperm and egg, a human brain could grow into the most complex thing in the universe, with billions and billions of nerve cells making trillions of connections, here's what she told me: "It's almost overwhelming to think about the whole thing. If you think about how the whole brain and nervous system get assembled . . . you just want to throw up your hands and say, 'It's way too complicated.'"

Scientists explore enormous questions like mine, McConnell explained, by breaking them down into smaller ones, designed to yield incremental answers. My question, 'How does the brain build itself?,' translates into a series of investigations: How do neurons first form? How do they find their place in a growing brain? How do they make connections with other neurons? Then these questions are parsed into still smaller questions—anatomical, cellular, molecular—piling speculation upon speculation, experiment upon experiment, hypothesis upon hypothesis.

Suddenly I saw why science had always been so difficult for me. Lost in a thicket of proliferating questions, I was unable to find my way back to that initial curiosity and wonder that inspired the investigation in the first place.

Filming *The Secret Life of the Brain*

Here, then, was the first challenge for our documentary: to explore the various levels of complexity at which scientists work without losing sight of the fundamental questions about the human condition that are at stake.

I've always felt confident that television can communicate complex ideas as long as they are rooted in the emotional contours of our lives. For me, television works most effectively when thought resonates with feeling. Television may make information hard to digest, but not ideas. Too many facts interrupt the emotional flow of the narrative, threatening to sink it under the weight of endless detail. That's why I wanted our films to elaborate scientific ideas in the context of real-life human situations. If the stories were intense enough, people would remember them, and the concepts they embodied. I soon discovered that my instincts about television were supported by science: studies of memory have demonstrated that the ideas and facts we remember best are those charged with emotion, proving to me once again how surprising, and gratifying, making documentary films can be.

Our central concept, the theme that we would pursue in every program, emerged slowly. When we began research, exploring brain functions like memory, vision, and cognition, we were calling the series *Secrets of the Brain*. But with the help of science reporter June Kinoshita and the assistance of the Dana Alliance for Brain Initiatives, I soon learned that, more and more, neuroscientists were looking at the brain from the point of view of its development, studying the brain's plasticity—that is, its ability to change and grow over the course of a life-time. When we are young, our brains are at their most plastic, a real advantage during a period when we have to learn as quickly as we can in order to equip ourselves to survive. Contrary to what was once generally believed, however, the brain continues to evolve and change along with the rest of the body through-out our lives.

The brain and its capacity for change became the central theme of our series, and with it came a new title, *The Secret Life of the Brain*, with an emphasis on "life." I liked the feel of that. I imagined five programs—the brain of the baby, the child, the teenager, and the adult, and the brain in old age—each program framed by a question:

How does the brain form?

How does a child acquire language?

Is there a connection between brain development during adolescence, the onset of schizophrenia, and the prevalence of teenage addiction?

How does an adult find a balance between reason and emotion?

Why do some people remain energetic and vital in old age, while others do not?

The Secret Life of the Brain pursued these questions by telling stories about the brain's remarkable journey over the course of a lifetime. In our first program, for example, we tell how a baby, born with a cataract, needs to have the lens cloud-ing her vision removed as soon as possible because, in order to develop normally, the areas of the brain responsible for vision require the stimulation of electrical pulses generated by light striking the retina. If the brain cells devoted to vision do not form the proper connections with one another early on, they never will. Without an operation in the first few months of her life, the baby risks blindness in the clouded eye, even if she has the cataract removed later, and regardless of the subsequent health of the eye itself.

To my taste, when a single instance becomes emblematic of a larger concept, television is working at its best. The cataract story embodies a large idea: Our brains and the world around us are involved in a delicate duet; our brains change and adapt in response to an environment acting upon our genetic endowment. But emblem making can be misleading. It's easy to construe meanings too broadly —in this case, to extrapolate from visual development to the development of other parts of the brain. With the help of neuroscientists Pat Kuhl and Helen Neville, I learned that the brain consists of many different systems, each developing at its own pace and in its own way. Development, in other words, isn't etched in stone; the brain doesn't run on a deadline like a train leaving the station at an appointed hour. Our brains evolve gradually over a lifetime – which is good news for parents. Eager to encourage the intellectual and emotional development of their children, some parents worry obsessively about providing "developmentally correct stimulation" and earnestly turn to science for guidance. Because of one well-publicized but misconstrued study suggesting the beneficial effects of Mozart on the brain, for example, eighteenth century harmonies drift sweetly over the cribs of thousands of infants with no real evidence that music does anything for the baby other than to soothe a well-meaning parent's anxiety.

Fortunately, parents can relax. The first years of life are only the beginning of a slow process of growth that fathers and mothers can encourage by simply spending time enjoying their children. What once used to be referred to as "critical periods" of development, scientists now call "sensitive periods." We humans could hardly have survived as long as we have if our species were solely dependent upon specific experiences at specific times. As dramatic as the cataract story is, infancy and early childhood are the first stages of brain development, but not the last. It's like laying the foundation of a house, one scientist told me. Without the foundation, the house cannot stand, but construction doesn't stop there. During those first years, the infant brain develops very quickly: language, cognition, perception, and the major behavioral systems are put into place, but it is the fine-tuning of these systems throughout our lives that ultimately accounts for who we are.

Bringing good science to television is a complicated business. I was fortunate to work with a group of talented filmmakers who were comfortable with complex ideas and could examine them in all their human dimensions. Sarah Colt, Jenny Carchman, Ed Gray, Tom Jennings, Michael Penland, Amanda Pollack, and Annie Wong spent many months bringing the series to life, with Lesley Norman ably keeping the machinery of production running smoothly.

I was equally fortunate to work with Richard Restak, who has taken the concepts we developed for *The Secret Life of the Brain* and expanded and elaborated them into this clear and thoughtful book, traveling down pathways where our series had no time to go, and lingering in a way that only a good book can. In a film, nothing stands still. Our series takes you on a visual journey into the brain, down nerve fibers, across the microscopic synaptic gap between nerve cells, even into the cell itself, but Restak's book gives you the time to pause and reflect. Adding his own research to ours, Restak provides new layers of analysis and insight, giving an interested reader time to consider and re-consider. His book is the perfect complement to our film.

Now that *The Secret Life of the Brain* is done, what strikes me is not only how much scientists have discovered in just the last ten years, but how much there still is to know. Perhaps that's why, in making this series, I so often turned to poets to articulate the mysteries that remain. After all, Wordsworth called poetry "felt thought," or, put another way, "how an idea feels." Scientists and poets, I learned, share the same passions and puzzle over the same problems.

Pondering the immensity and power of the small, crinkled organ weighing less than three pounds, Emily Dickinson tells us that "The brain is wider than the sky," and more, that "The brain is just the weight of God." Theodore Roethke, attempting to fathom the delicate relationship between cognition and emotion, writes, "We think by feeling," anticipating the research of neuroscientist Antonio Damasio and Joseph LeDoux, among others. And poet laureate Stanley Kunitz speaks directly to the theme of *The Secret Life of the Brain* when he writes:

> "I've walked through many lives,
> some of them my own,
> and I'm not who I was,
> though some principle of being
> abides, from which I struggle
> not to stray..."

At 95 years old, Kunitz is himself a testament to the brain's remarkable ability to learn and adapt, even into old age:

> "... no doubt the next chapter
> in my book of transformations
> is already written.
> I am not done with my changes."

David Grubin

INTRODUCTION

Let's imagine for a moment a family—call them the Kellys—where an argument is taking place between Shawn, 18 years old, and his father, Michael. In attendance are Shawn's six-month-old sister Katie, lying peacefully (until now) in her bassinet; Shawn's six-year-old brother Kevin, who divides his attention between the argument and a baseball game he's watching on television; and Shawn's grandfather, who look increasingly uncomfortable as the argument progresses. The argument concerns Shawn's unwillingness to return home with the family car by midnight. What "world" is each participant perceiving?

Katie doesn't perceive much of anything except loud noises coming from farther away than she can see clearly. Although her brain enables her to recognize human speech sounds, her vocabulary is zilch and she hasn't a clue about cars, curfews, or adolescent rebellion.

Kevin experiences the argument as only a distraction from the game that he's trying to watch. And distractions are very much a part of Kevin's life. He has trouble sitting in one place for more than a few minutes or keeping his mind focused. The shouting in the background only increases his difficulties; if asked, he couldn't say what the argument is about.

For Shawn, important issues are at stake. He's "sick and tired" of being told what to do, not trusted, treated "like a child." It's time Dad starting realizing that he's no longer a six-year-old like Kevin.

Shawn's father also sees important issues. Is his kid going to conform to household rules that everyone understands, or is he going to consider himself "better than everybody else" and no longer obey his parents?

At age 82, Grandfather can look back on similar arguments he's seen and heard on many occasions. "Perfectly normal adolescent rebellion based on

Shawn's need to establish identity and boundaries," he mutters to himself. If Shawn and his father could only see each other's point of view for just a moment, he thinks.

Each participant in this minidrama perceives the situation based on the maturational development of his or her brain. In Katie's case, brain development hasn't proceeded sufficiently for her to hold any point of view whatsoever. At the other end of the brain-maturational scheme, grandfather has a lifetime's worth of observations of the interactions of adolescents and their parents. He recognizes the futility and counterproductiveness of such spats, thanks to the wisdom provided him by the normal functioning of a part of his brain called the prefrontal cortex. That most-developed and elaborated portion of his brain provides him with the confidence that everything can be resolved if each person has the wisdom to put himself in the other's place.

In short, a "family argument" isn't the same for each participant. More important, the differences could be altered by suitable alterations in the brains of the participants. If Kevin were a few years older, he could understand why Shawn feels so strongly about asserting himself. If Shawn were his father's age he might be curious about what's happening in his father's life and whether this display of parental authority has anything to do with the problems his father has been encountering recently at work. If Shawn's father could exert firmer control over the emotional centers in his brain, he might be able to identify with how he used to feel not so many years ago when he was Shawn's age. While performing this imaginative exercise, the reasoning powers of his prefrontal cortex would be proving to him the futility of the argument while simultaneously suggesting compromises. Only grandfather sees the situation for what it is: an example of intergenerational conflict that is best handled by stepping back, seeing the other person's point of view, seeking a compromise, and most of all, placing the situation in context.

Notice that while each of the suggested behavioral modifications can be described in psychological terms ("seeing the other's point of view," looking for "compromises," et cetera), the underlying processes depend on the maturation of the brain. And this changes across the developmental spectrum from infancy to childhood to adolescence to adulthood to old age.

In *The Secret Life of the Brain* we explore the five developmental stages in the human life-span and their implications for health and happiness. Development, as we will see, is a two-edged sword. While it provides opportunities for happiness and achievement, it can also exert destructive effects if the developmental sequences go wrong. For instance, the same search for identity and affirmation that nudges Shawn into adolescent disagreement with his father can also lead him into unhealthy relationships with peers involving drug use or impulsive acts like careless driving (perhaps one of the unstated fears of Shawn's father).

Each of the brain's developmental stages provides its own opportunities and perils. Each is part of a marvelous narrative that starts at the moment of conception and extends to the last breath. And as with every good story, it's best told by starting at the beginning. Here's a short overview of what we'll be exploring:

If we were to employ a single word to characterize the human brain in all stages of development it would be *plasticity*: the organ's capacity to change. The brain's plasticity distinguishes it from anything else in the known universe. Without plasticity, the brain would be incapable of adjusting to changing times and conditions. Indeed, absent its plasticity, the brain would be similar to a machine, a structure with strictly limited powers of adaptation to the environment. Instead, thanks to plasticity, the brain possesses amazing powers of adaptation and recovery. Throughout this book the theme of plasticity can be discerned as a subtext to everything the brain does. Even as you read these words, your brain is changing as a consequence of your brain's plasticity.

Plasticity is most evident during infancy. From a tiny ball of cells the brain first emerges, grows, and organizes itself. As the brain increases in size and complexity inside the womb, its growing cells interact with their environment and with one another. Newly formed neurons establish connections and those connections multiply throughout infancy.

Experience provides the basis for the formation of the connections and the transformation of those connections into circuits. Change the experience and you change the brain. Deprive the baby's brain of light and sound and human contact, and it will remain stunted. The same thing will happen if the brain enters the world too soon and in its prematurity is overwhelmed with more stimuli than it's equipped to handle.

But plasticity doesn't stop at infancy. It continues across the entire life-span: It is present in the infant's brain; the brain of the infant's five-year-old sibling who has just learned that Mother will be bringing home "a sister" from the hospital; the brain of the adolescent brother as he stares down at his new sister; the brain of the proud father in the delivery room getting his first look at his child; and the brain of grandmother as she sits proudly cuddling her new grandchild. At all of these stages and at every age the brain retains its capacity to change in response to life experiences.

And plasticity takes different forms during the five stages of human development. Although we will discuss the stages in greater detail in the following chapters, here is a thumbnail sketch of what takes place.

During *gestation* the brain makes its first appearance as a crest of cells from which the neurons, brain cells, will emerge. Next come the formation of the major brain regions and a migration of neurons from their original sites of generation to their final positions in the brain. As we will see, disastrous consequences can ensue whenever one of these processes goes awry as a result of disease, genetic mutation, or exposure to drugs or chemical toxins. Indeed most congenital (present at birth) brain defects result from disruption of the normal programs of neuronal growth, development, and migration.

During *childhood*, the second life stage, a new kind of plasticity takes precedence. With most of the neurons in place at the conclusion of infancy, a sculpting process emerges as the dominant force shaping the brain. At this point the brain contains many more neurons than it requires, and excess neurons must be pruned away according to the most fundamental tenet of brain operation: Use it or lose it. In practical terms, this means that the overabundance of neuronal connections established during gestation and infancy is thinned out in response to experience. Unused or rarely used pathways disappear, while heavily trafficked pathways flourish and elaborate. Thanks to this process of forming, re-forming, and strengthening neuronal connections, young children proceed by leaps and bounds in their abilities to pay attention, remember, and make their first efforts at mastering the universe.

In *adolescence* the pruning process has largely taken place in most areas of the brain, with one notable exception: the prefrontal cortex. Select any of the difficulties associated with adolescence (impulsiveness, erratic mood swings, rebellion against authority, poor judgment, et cetera) and you'll find that those difficulties are the result of immaturity in the prefrontal cortex. During adolescence pruning occurs within the immature prefrontal cortex and elsewhere. While the pruning is taking place, the adolescent remains "difficult," "unpredictable," "moody"—fill in your own favorite adjectives based on your experiences with adolescents. But when that pruning has been successfully completed we employ a different set of adjectives: The adolescent is now "more mature," "thoughtful," "likable," and even "courteous."

By *adulthood* all the brain areas are up and functioning. The brain contains its full repertoire of cells, although some new cells may be added, principally in discrete brain areas such as those associated with memory. Adulthood is the culmination of human brain development, the goal that nature was striving for. More than half of all the brain cells lie within the cerebral hemispheres, crowned by the wrinkled surface area called the cortex. Hidden within its folds reside our powers of thought and reason.

Old age was once thought of as principally a holding operation: retaining past gains while yielding as little as possible to the ravages of age and disease. We now know that such a view is unduly pessimistic. The healthy brain of the aged person retains a marvelous plasticity and can change for the better and the worse. Indeed, the brain of the older person can continue to function healthily and creatively, or it can decline in its powers due to abuse or simple lack of use. The older person faces a choice: accept the stereotypes about aging and sit in a corner, or remain active and vibrant, perhaps even becoming like writer Harriet Doerr, who wrote her first book, *Stones for Ibarra*, while in her 80's. But in order to thrive during this last stage of life, dangers must be anticipated and protected against. For instance, late-life-onset alcoholism is frighteningly common in people in their 70's and 80's who fail to keep their brain finely honed by reading and other intellectual pursuits. Again, as in the other stages of brain development, we encounter that double-edged sword of plasticity.

When things go wrong in the brain's growth or development, a smorgasbord of human afflictions and pathological conditions can result, such as drug abuse, schizophrenia, and depression, and even suicide, that most sorrowful of calamities that can befall the brain. But in the vast majority of instances, the brain develops perfectly normally. In practical terms, this means that at one or another of the five stages of development the brain can bring comfort, excitement, insights, and a host of pleasures.

While the brain is but one organ among many in the human body, it is the source and determiner of everything. Our understanding of the world changes in concert with the evolution of this delicate structure, which is unlike anything in the universe. Indeed, we understand the world the way we do at each of life's stages because of our brain. And yet, until lately, the brain jealously guarded its secrets. Only recently—with the development of powerful technologies—have we been successful in delving into the secrets of the brain.

How We View the Brain

The brain's physical construction and location contribute to its "secrecy." Encased within the bony confines of the skull and enveloped by three membranes, the brain is the most elusive organ in the body. To observe it directly, the outer two of its coverings (the skull and the underlying dura mater) must be broached. During a craniotomy (a neurosurgical operation on the brain), the neurosurgeon employees a high-speed drill to cut an opening through the skull. He then

retracts the bone, cuts through and lays open the thick, leather-like layer of the dura. At this point the brain comes into direct observation. But since a craniotomy entails risks of danger, and even death, the operation could never be justified simply to open up the skull of a normal volunteer and "take a peek inside to see what's going on in there." For this reason, the lion's share of our knowledge about the brain awaited technological advances capable of providing indirect views of brain *structure* (how the brain is built, its anatomy) and *function* (how the brain operates).

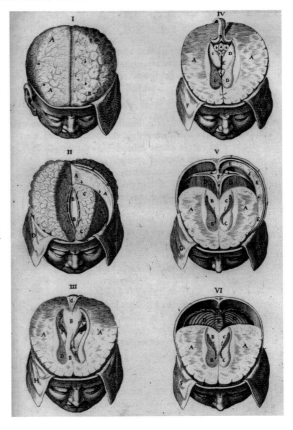

Historically, structure-revealing technologies preceded functional ones. The advent of X-ray technology first made possible a limited view of brain structure. But since X rays provide only a two-dimensional image of a three-dimensional object, X-ray pictures aren't much help in

Early efforts to learn about the brain are captured in this series of 17th-century illustrations by Johannes Vesling, a professor of anatomy and surgery in Padua, Italy. Vesling's drawings, depicting the layers that lie beneath the skull, were praised as "inferior to none in the world."

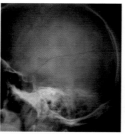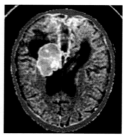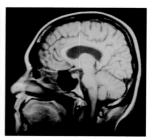

With the advent of X rays (left), researchers could see beneath the scalp without surgery. By combining a series of X rays with the power of the computer, CT scans (middle) can locate brain tumors and other structural abnormalities. Magnetic resonance imaging (MRI) uses the magnetic qualities of the brain's atoms to offer greater resolution of the various areas of brain structure (right); tissues with high concentrations of water appear lighter, bone appears darker.

revealing the brain's secrets. Furthermore, interpretation is ambiguous and unreliable since the images overlap. A CT scanner (short for computerized axial tomogram) gets around these difficulties and provides an easily readable three-dimensional picture.

CT scans use a moveable X-ray tube that revolves around the patient's head as it lies within a doughnut-like configuration of sensitive X-ray detectors. As the X-ray beam rotates from detector to detector, a computer converts the assembled information into a series of images. And since the X-ray beam moves sequentially around the entire circumference of the head, it gathers information from the entire brain. Visual "slices" of the brain (*tomos* is Greek for "section") can then be generated through various planes and the corresponding images displayed on a screen.

A second imaging technique, magnetic resonance imaging, or MRI, relies on the brain's selective absorption of radio waves rather than X-rays. When atoms of certain elements like hydrogen are placed in a magnetic field, their nuclei behave like tiny bar magnets, aligning themselves with the magnetic field. Then a pulsed radio wave is turned on and knocks the atoms out of alignment. When the pulse is abruptly turned off, the atoms return to the original alignment imposed by the magnetic field. In the process of this realignment, the atoms emit detectable radio signals that can be collected, collated by a computer, and converted into detailed brain images.

MRI provides greater optical resolution than CT (less than a millimeter, compared to several millimeters for CT). But while both techniques provide detailed images of brain structure, neither technique provides a means for observing brain functioning. Using CT or MRI is like looking at the floor plan and seating arrangement of a theater: They may provide you with an idea of where you'd like to sit, but they don't enable you to see the play or hear the actors.

A variant of MRI called functional MRI (fMRI) provides you with the equivalent of the movement and dialogue of the actors in the play. This imaging technique takes advantage of the fact that the magnetic properties of blood change according to the amount of oxygen carried in the blood. Specifically, hemoglobin, the oxygen-carrying molecule in blood, emits different signals depending on whether it is oxygen-rich or oxygen-deprived. When a certain part of the brain is active, more oxygen is required and, as a result, increased oxygen-rich hemoglobin is delivered to the area. An fMRI scan records and translates this increase in oxygen-rich hemoglobin into computer images. Reading this sentence, for example, increases the amount of oxygen-rich hemoglobin to the areas of your brain responsible for vision. An fMRI would show enhanced activity in your occipital lobes as a reflection of this increase in oxygenated hemoglobin.

Scan by PET (positron emission tomography) also provides computer-generated images based on brain activity. In this technique, the subject is injected with a radioactive isotope (usually linked with glucose) and its decay sequence is measured and converted into images. PET makes possible the localization of specific mental operations.

Both fMRI and PET provide images of functional brain activity—that is, the dialogue and actions of the play rather than just the seating structure of the theater. But even these highly sophisticated technologies are limited to measurements of only three of the brain's metabolic changes: blood flow, oxygen use, and glucose consumption. Neither PET nor fMRI provides any information about the

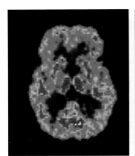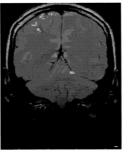

With PET, a radioactive substance attached to glucose molecules gives off gamma rays that are translated into color images; at far left, active areas of the brain appear red. Functional MRI (fMRI) captures the difference in magnetic signals between oxygen-poor blood and the oxygen-rich blood that signals heightened brain activity (red and yellow, near left). Operating at the rate of one image per second, fMRI enables researchers to determine whether activity occurs simultaneously or sequentially in different regions of the brain as a patient reacts to experimental conditions.

Electroencephalograms (EEGs, below left) show researchers the brain's electrical activity in real time. Magnetoencephalography (MEG) uses sensors on the skull to record the magnetic flux associated with electrical currents in activated neurons, in this case during a subject's response to a brief tone. MEG results can take different forms: The head plot (middle) shows the magnetic field change in each MEG sensor; the contour map (below right) shows the magnetic field distribution as a signal consisting of two "dipolar" sources. MEG scans can localize sources in the brain to within a few millimeters, more precise than the one centimeter resolution typical of EEGs.

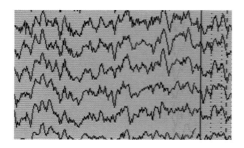

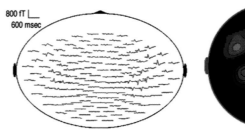

800 fT
600 msec

electrical fluctuations between neurons that also characterize brain activity. In order to monitor electrical events, electrical recording devices must be used, such as electroencephalograms (EEGs). An additional refinement employs computers to "filter out" extraneous electrical background "noise" so that the resulting EEG is focused more precisely on selected electrical brain activity.

Finally, magnetoencephalography, or MEG, works by recording the magnetic fields produced by the living brain. These fields are extremely small, almost a billion-fold smaller than the ambient fields produced by the earth and many other magnetic sources in our environment. Nevertheless, MEG can localize magnetic field changes to the nearest cubic millimeter of cerebral cortex. It can also time any changes occurring within the brain to the nearest thousandth of a second.

Each of these different imaging techniques—CT, MRI, fMRI, EEG, and MEG—provide a different window on brain structure or function. While CT, MRI, fMRI, and PET achieve excellent spatial resolution, only EEG and MEG can provide measurements of brain activity at resolutions approaching "real time." As a result of these differences, each investigator will choose an imaging technique best suited to his or her interests. A neurosurgeon suspecting a brain tumor, for instance, will initially employ a CT or MRI in order to detect any deviations from normal brain structure. A researcher investigating the brain areas involved in reading or speech will choose an fMRI or PET in order to observe dynamic activity patterns within different parts of the brain as the subject reads or speaks. It's likely that future technological advances, especially new imaging methods, will provide not only new information about the brain but also far-ranging and innovative ways of studying it.

The brain is wider than the sky,
   For, put them side by side,
The one the other will contain
   With ease, and you beside.

The brain is deeper than the sea,
   For, hold them, blue to blue,
The one the other will absorb,
   As sponges, buckets do.

The brain is just the weight of God,
   For, lift them, pound for pound,
And they will differ, if they do,
   As syllable from sound.

Emily Dickinson

Wider than the Sky

THE BABY'S BRAIN

A baby's brain weighs less than a pound. Yet from within its tiny folds will emerge a universe of meaning: emotions, ideas, memories, dreams—indeed, everything that makes us human will find a home here. At the moment of conception the brain is secret, indeed. All that can be observed under a microscope is the single cell that resulted from the penetration of the father's sperm into the mother's egg. But within that tiny cell, invisible to the naked eye, resides the DNA blueprint that will guide the construction of the entire human body. And that construction gets off to an early start.

After an initial series of rapid multiplications, the resulting cells—many hundreds —arrange themselves into a hollow sphere surrounding a central cavity. The formation of this hollow sphere marks life's first developmental step.

Next, at about two weeks after sperm meets egg, the embryo undergoes a massive rearrangement that transforms a uniform ball of cells into a multicelled organism with a recognizable body plan. This process of differentiation starts with an indentation of the cellular sphere. A portion of the sphere then moves inward through the indentation, resulting in a three-layered structure: an outside layer (the ectoderm), a middle layer (the mesoderm), and an inner layer (the endoderm). From these three layers will emerge distinct bodily structures.

Development of the Brain During Gestation

The formation and development of the human brain is the most complex job in the world. At birth, the baby's brain weighs less than a pound. Yet the brain cells, or neurons, have formed more connections than there are stars in the universe.

By three weeks of development (enlarged to show detail; the actual size of the embryo at this stage is about that of a grain of rice), neurons are forming at a rate of more then 250,000 per minute. The final number of neurons at birth will surpass one hundred billion.

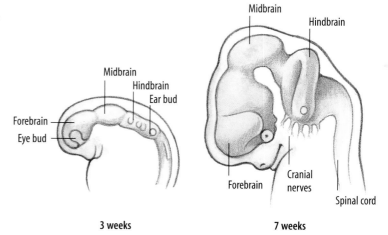

3 weeks

7 weeks

The endoderm gives rise to the gut and many of its major organs; the mesoderm forms the body's muscles, skeleton, connective tissue, the heart and circulatory system, and the urinary tract and genitals; the ectoderm forms the skin and the brain, along with the rest of the nervous system.

The future brain and nervous system first become apparent at about four weeks, when a portion of the outer ectoderm thickens to form a spoon-shaped structure only one cell thick known as the neural plate. A groove (the neural groove) runs the length of the neural plate, dividing it into right and left halves.

Defining Characteristics Even at this early stage of development the future brain possesses three defining characteristics. It is polarized (the head end is wider than the remainder of the neural plate), bilaterally symmetrical (divided into right and left halves separated by the neural groove), and regionalized (the wide end of the spoon will become the brain, while the narrow end will develop into the spinal cord).

Next, the two sides of the neural plate fuse to form a tube from which emerge three swellings: the forebrain, midbrain, and hindbrain. Over the ensuing months in the womb these three swellings enlarge, bend, and expand to form the major divisions of the adult nervous system: from top down the cerebrum, the midbrain, the thalamus and hypothalamus, the cerebellum, and the spinal cord.

Scientists who study the brain during its earliest period marvel at the clockwork precision by which the genes issue instructions for growth and development.

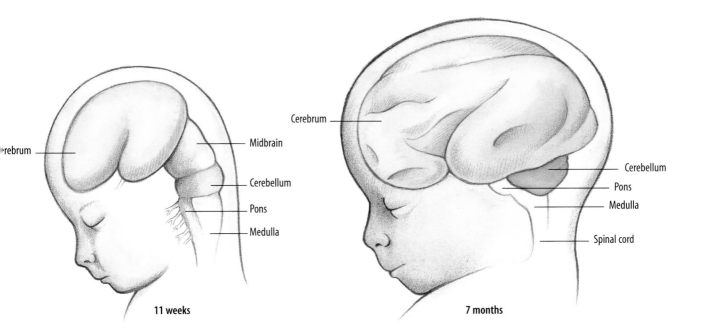

rebrum

Midbrain

Cerebellum

Pons

Medulla

11 weeks

Cerebrum

Cerebellum

Pons

Medulla

Spinal cord

7 months

As development progresses, each neuron will make as many as 10,000 connections to other nerve cells in the brain. Amazingly, unlike other cells in the body that regularly die and are replaced, the neurons you have in old age are the same neurons that formed during development in the womb.

In the earliest weeks of brain growth, the forebrain, midbrain, and hindbrain appear as three enlargements along a tubelike structure (above, far left). By seven weeks the forebrain has enlarged out of proportion to the other two. It will ultimately give rise to the cerebral hemispheres. The midbrain will give rise to the adult midbrain, and the hindbrain to the cerebellum and the lower brain stem (pons and medulla). Midbrain and hindbrain structures are soon hidden from view by the cerebral hemispheres, which at about seven months (above, far right) take on a wrinkled appearance. While sharing common features, each brain's developmental pattern is unique. As a result, no two brains are exactly alike.

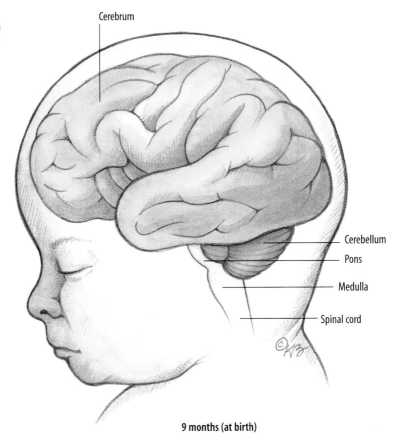

Cerebrum

Cerebellum

Pons

Medulla

Spinal cord

9 months (at birth)

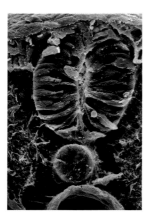

Within the embryo, chemical cues guide the development of neural cells such as the developing spinal cord cells seen in this scanning electron micrograph. The cells make up the just emerging neural tube (large oval). Interactions between the neural tube and the notochord (smaller round shape) help spur development of motor neurons in the adjacent portion of the neural tube.

But even as it begins to form, the brain remains highly dynamic. And the environment of the embryo plays a crucial role in how the brain will finally turn out.

According to neuroscientist Mary Beth Hatten, a developmental biologist at Rockefeller University, in New York, "The embryo itself provides the environment that interacts with the genes. It's teeming with chemicals, and its very shape has an influence on the brain. And as the brain grows it also creates an environment that interacts with the genes and can lead to changes in how the brain will eventually develop. Thus the brain is a dynamic organ right from the beginning."

Dramatic Growth of the Cerebral Hemispheres   A glance at the illustrations on pages 2 and 3 reveals the brain's most dramatic developmental alteration over time: the exceptional growth of the forebrain, which gives rise to the cerebral hemispheres. When the brain is viewed from the side, only three of its major structures are visible: the brain stem, the cerebellum, and the cerebral hemispheres. All other structures are hidden by the vastly expanded cerebral hemispheres, which represent 85 percent of the brain by weight.

In addition to their large size, the hemispheres are remarkable for their highly wrinkled, convoluted appearance. At five months of age the cerebral hemispheres appear as smooth as billiard balls. Four months later they look more like the two halves of a gnarled walnut.

"If one looks at the brain of the developing infant in utero, it's really quite smooth, but during the last 12 weeks of prenatal development the brain folds in on itself," Michael Rivkin of the Department of Neurology at Children's Hospital, in Boston, told us. "By the end of the pregnancy the surface of the infant brain consists of a landscape of hills and valleys, technically referred to as gyri and sulci. By the time the infant is born, its brain looks like the brain of an older child or young adult."

Why does the brain fold in on itself and undergo such a dramatic change in appearance? Think of the last time you packed a suitcase. Folding your clothes allowed you to enclose the comparatively large surface of your wardrobe within the fixed confines of the suitcase. A similar situation exists in the brain: A large surface area can be crammed into the fixed volume of the human skull only by wrinkling and enfolding. "The surface of the brain folds in on itself as a way of

Throughout gestation, the brain increases in size and the cortex grows increasingly wrinkled as the connections between neurons strengthen and weaken. New experiences in the first week after birth (far right) cause even more bulges and furrows. All these hills and valleys increase the surface area of the cortex, so to stay within the confines of the human skull, the cortex folds in on itself many times over. If the cortex were smoothed out, we would need a skull the size of an elephant's to accommodate it.

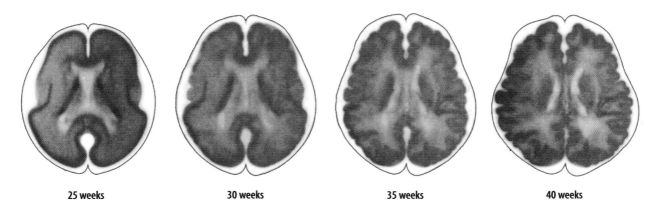

25 weeks   30 weeks   35 weeks   40 weeks

accommodating an increasing surface area without changing the intracranial volume into which it has to fit," says Rivkin.

A large surface area is important because it increases the number of neurons (brain cells) that can be accommodated within the cerebral cortex, the outer 2 millimeters of the hemispheres. And the cerebral cortex represents perhaps nature's most striking example of Mies van der Rohe's famous aphorism, "Less is more." This thin rind (*cortex* means "rind" in Latin) that is less than the thickness of an orange peel and has the consistency of tapioca pudding contains two-thirds of all of the 100 billion neurons in the human brain and almost three-quarters of the 100 trillion interneuronal connections.

The Amazing Cerebral Cortex   Making up seven-tenths of the entire nervous system, the human cortex is 10 times larger than that of a macaque monkey and 1,000 times that of a rat. Indeed, neuroscientists are convinced that it's the cerebral cortex that sets us apart from all other creatures. And that increase in cortical size includes a tremendous increase in the number of brain cells. Even a piece of human cortex the size of a grain of rice contains about 10,000 of them.

Since the layer of neurons consisting of the cortical cells is gray in appearance when looked at with the naked eye, early students of the human brain referred to it as *gray* matter. In contrast, the layers located beneath the cortical layer appear

The Crown of Creation

Every aspect of human life—from basics such as breathing to the expression of personality—is governed by the brain. Each of the brain's many components performs its own tasks, but these components also work together.

The medulla, pons, and midbrain, which together constitute the brain stem, handle critical body functions that proceed without conscious effort—regulating heartbeat and breathing, for example. Above the brain stem lie the various components of the limbic system, which take care of a host of basic brain activities, including the physical expression of emotion and the processing of sensory information.

Almost completely enveloping these structures are the two hemispheres of the cerebrum (left). The cerebrum's most significant feature is its thin outer layer, the cerebral cortex, which consists of four lobes in each hemisphere (below). Working with the brain's other components, this distinctively human part of the brain produces thoughts, governs language, and stores memories.

LOBES OF THE CEREBRAL CORTEX

Each of the four lobes is a specialist. The occipital lobes are responsible for vision; the parietal lobes for the analysis of sensation; the temporal lobes for hearing, understanding speech, and the formation of an integrated sense of self; and the frontal lobes for carrying out executive functions such as decision making and foreseeing the consequences of one's actions. Despite this specialization of the different lobes, the brain functions as a unit and integrates the contributions of the different lobes into a single unified experience.

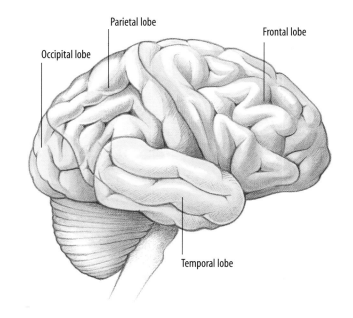

Occipital lobe

Parietal lobe

Frontal lobe

Temporal lobe

A view of one-half of the brain reveals some of the many structures that govern our daily lives. The hippocampus, for example, is involved in memory formation. The amygdala and fornix play important roles in emotions. The expression of the endocrine and autonomic nervous systems are the responsibility of the pituitary and hypothalamus, while the thalamus handles sensation. Key to the integration of the two cerebral hemispheres is the corpus callosum. Below these structures, the midbrain and pons control the nerves responsible for eye movements, the size of the pupils, facial sensation and movement, hearing, and wakefulness. The cerebellum receives input from many parts of the nervous system and sends signals to those parts important for the control of movement. The medulla contains centers for the control of breathing, swallowing, and blood pressure, among other vital functions. Below the medulla is the spinal cord, which conveys motor signals from the brain to the rest of the body and carries sensory information from the body to the brain.

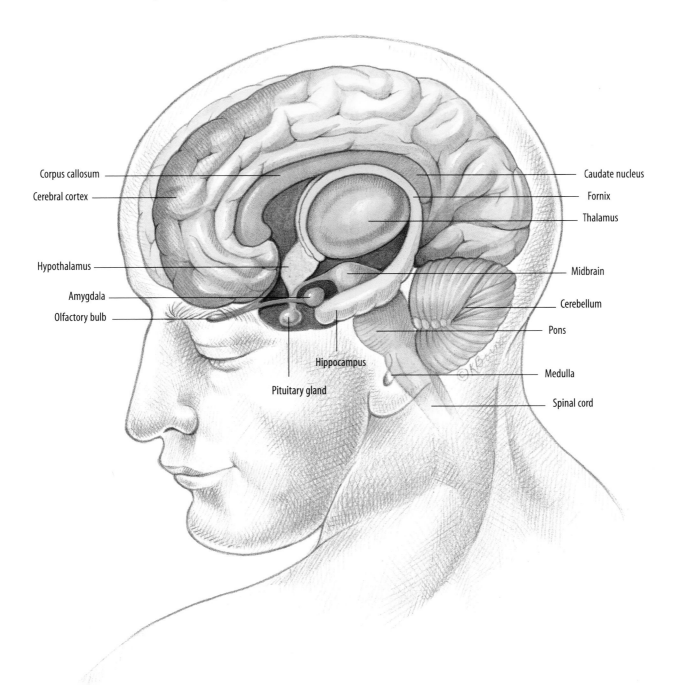

Corpus callosum

Cerebral cortex

Hypothalamus

Amygdala

Olfactory bulb

Pituitary gland

Hippocampus

Caudate nucleus

Fornix

Thalamus

Midbrain

Cerebellum

Pons

Medulla

Spinal cord

white when observed by the anatomist. This *white* matter consists of the communicating filaments—called axons and dendrites—that enable one nerve cell to communicate with another, whereas the gray matter is the home of the brain cells.

The linkage between intelligence and "gray matter" is firmly fixed in our language. For instance, the fictional Belgian detective Hercule Poirot frequently spoke of employing the "little gray cells" in order to reason his way toward the solution of a crime. And a common disparaging expression used for people given to doing not very smart things is that they lack "gray matter."

But intelligence is not just the result of the raw number of neurons composing the gray matter in a particular brain. If this were true, then adults with larger heads and brains would be smarter than those with smaller heads and more modestly sized brains. Rather, intelligence has more to do with the number of interconnections among neurons. And the size and complexity of this vast network of interconnections varies at different stages of life and according to the challenges that the brain encounters. When fully developed, the human brain's 100 trillion neuronal connections exceed those of any other creature on earth. In addition, as we will discuss in more detail later, the brain retains the potential to establish new connections throughout our lives.

An Axon's Reach   In order to establish its connections, each neuron possesses a single fiber, called an axon, that stretches away from the cell body and makes contact with neurons elsewhere in the brain. Some axons reach only a few millimeters to influence nearby neurons, while others cross over from one side of the brain to the other and thereby influence neurons several centimeters away. Or they even wend their way a meter or more (axons stretching from the human spinal cord to the foot are about a meter in length and at their terminal point perform the humble but necessary service of activating muscle cells moving the big toe).

The connections of one neuron with another via the axon take place on specialized receiving areas known as dendrites. (A smaller number of axons directly influence the receiving neuron's cell body.) While each neuron possesses only one axon, it may possess thousands of dendrites. This arrangement makes it possible for a neuron to exert a widespread influence within the brain: Each neuron's single axon contacts the dendrites of thousands of other neurons.

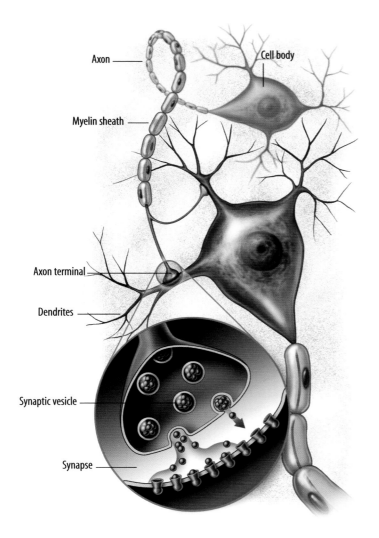

Axon

Cell body

Myelin sheath

Axon terminal

Dendrites

Synaptic vesicle

Synapse

Neurons communicate with one another through a series of electrical and chemical signals. The electrical impulse begins where a fiber called the axon leaves the cell body. It travels the length of the axon, moving across gaps in the fatty sheath of insulating myelin, which speeds the signal along at rates ranging from 9 to 400 feet per second. At the axon terminals, chemical transmission takes over. Structures called synaptic vesicles release molecules of a chemical neurotransmitter into the synapse, the minute gap between neurons. The molecules bind to receptors on the surface of the target neuron. If the message to the receptors exceeds an excitatory threshold, the target neuron will start its own electrical impulse to move the signal onward.

"You can think of the neuron as an old-style telephone," says Carla Shatz of the Neurobiology Department at Harvard Medical School. "The cell body of the nerve cell is like the base of the telephone and contains most of the operative machinery. The receiver of the telephone corresponds to the many dendrites that gather the information. The cell body integrates that information and then sends the result out along the long process of the axon, which is like the wire that connects one person's phone with other phones."

Connecting Across the Synapse   Connection, incidentally, doesn't imply a fusion of one neuron with another. The axon of the transmitting neuron is always separated from the dendrites of the recipient neuron by a tiny space called the synapse. This arrangement allows for maximal flexibility. The neurons are not physically tethered to each other, so neuronal interactions can form and

Even a small segment of the brain contains a multitude of neurons and the connections between them, as depicted in the artist's conception above. Two bright lights represent electrical impulses traveling the length of an axon to deliver a message from one neuron to another.

reform, varying from moment to moment. This instance of the brain's plasticity—our book's underlying theme—is further enhanced by the nature of information transmission among neurons.

From Electrical to Chemical and Back Again The transmission between neurons is both electrical and chemical. Initially a tiny electrical discharge is transmitted along the length of the transmitter axon until it arrives at the synaptic space. The process of information transfer then shifts from electricity to chemistry. A messenger chemical, the neurotransmitter, is released from the transmitter neuron and ferried across the synapse where it links up with the neuron at its special docking station dubbed, appropriately enough, as a receptor. After stimulating the receptor nerve cell into action (or quieting it if the interaction of neurotransmitter and receptor leads to inhibition rather than excitation), nerve cell communication reverts once again to electrical transmission. The electrical impulse travels the length of the axon until it reaches the synapse, where chemical processing takes over once again.

Since electrical transmission is exceedingly slow in terms of the brain's requirements for speed, during early development nerve cells generate an insulating fatty layer known as myelin. Myelin works as an insulator that speeds up nerve cell conduction during the electrical phase of transmission. The myelin works like the insulation wrapped around a telephone wire. The nerve impulse is transmitted more rapidly and efficiently than occurs in nerve cells lacking the myelin covering.

With only a few specialized exceptions, the entire neuronal population of the human brain is produced prior to birth from a small number of precursor cells located in the innermost layer of the neural tube called the ventricular zone. These precursor cells migrate and multiply at an astonishing rate. About 250,000 new neurons are generated each minute during the peak of cell proliferation. The eventual result of this prodigious multiplication will be the arrangement within the cerebral cortex of at least several dozen neuronal cell types that can differ in terms of shape, size, the configuration of cell surface molecules, and the identity of neurotransmitters employed, as well as the kinds of synapses developed. Indeed, without the billions of cells of the cerebral cortex there would be no axons to connect with the rest of the brain, no information transmission and exchange. Minus the cerebral cortex our mental lives would simply never come into existence.

To help conceptualize the process of neuron cell proliferation and migration, neuroscientist Mary Beth Hatten employs the homely analogy of a garden hose. "From the beginning, the brain has this remarkable geometry. First, there's a sheet of cells that rolls up into a tube. And then the tube just folds up, as if you took your garden hose and put some flexes in it. At this point the newly formed neurons migrate from the inner layer outward toward the outer wall of the hose. And they migrate in vast numbers. At any given time millions of cells are on their way."

Mass Migration   Imagine the drama of millions of cells moving outward from their point of origin in order to form the all-important neurons of the cerebral cortex. Along their journey they must somehow traverse outward the several millimeters from the inner ventricular surface to just slightly below the outer surface. And although the distances traveled vary greatly from neuron to neuron, most developing neurons migrate substantial distances.

"In the human brain the cells may travel several centimeters, which corresponds to you or I walking to California," Hatten told us. But the neurons don't just head out without maps or guides. They are aided in their trek to their eventual destination by specialized cells, the radial glia (*glia* means "glue"), that span the entire

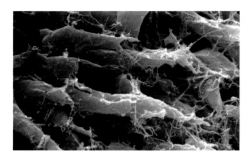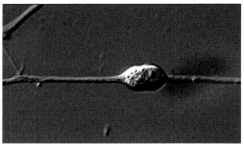

A multitude of neural crest cells are caught in the act of migration in the scanning electron micrograph at right. As seen in the view at far right, neurons travel along supporting structures called glial cells, some traveling several centimeters —the equivalent of a human being walking from New York to California.

thickness of the neural tube. The neurons crawl along the radial glial cells like snakes or inchworms entwining and extending themselves along a pole. What's more, the neurons of the six layers of the cortex arrange themselves in an "inside-out" manner. The neurons first on the scene are eventually located in the deepest layers of the brain, while later arrivals travel a greater distance and wend their way past the earlier arrivals until they reach their final destination closer to the brain's surface.

"It's like the westward expansion in the United States," according to Dr. Hatten. "The first immigrants came to New York and later immigrants pressed further forward to western New York, then Chicago, Denver, and finally on to California." As a result of this inside-out arrangement, each layer of the cortex— marked by distinct patterns of cell types and cell connections—is populated by neurons generated at different periods of development. And each neuron knows its ultimate destination.

"Nerve cells acquire information about their destinations about the time that the nerve cell is generated. From birth, the nerve seems to know where to migrate," according to Susan McConnell, a researcher at the Department of Biological Sciences at Stanford University. "Every individual neuron obeys a set of instructions as if it is following some sort of blueprint for development laid down in the genetic code."

Heredity or Environment?   When in this process does a brain cell know what type of cell it will be? Does the brain cell follow the British plan of Victorian England, in which heredity—who your parents are—plays the major role in identity, or does it follow an American plan, in which the environment— your neighborhood, schools, friends—is the major determiner?

Division of Labor in the Cortex

The structure of cells in the cortex varies from one region to another, giving each area of the cortex responsibility for different functions, as indicated below. However, no single part of the brain acts strictly independently; rather, the different regions communicate through a complex network. For example, impulses from the eyes, ears, and other parts of the body are received first by the visual, auditory, and sensory cortices. Higher-order regions, such as the posterior parietal cortex, analyze the signals for more information. Then the association cortices—the seat of perception and thought—integrate the sensory details into a coherent picture.

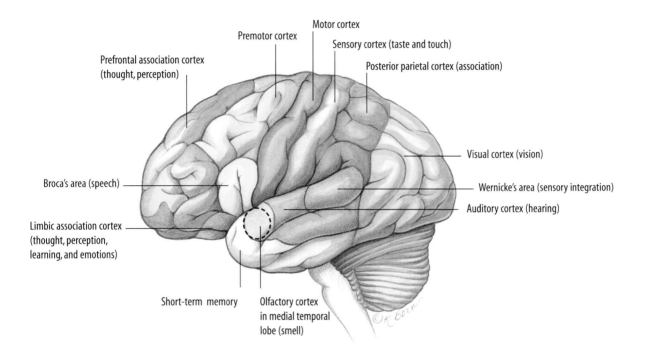

Premotor cortex
Motor cortex
Sensory cortex (taste and touch)
Posterior parietal cortex (association)
Prefrontal association cortex (thought, perception)
Broca's area (speech)
Limbic association cortex (thought, perception, learning, and emotions)
Short-term memory
Olfactory cortex in medial temporal lobe (smell)
Visual cortex (vision)
Wernicke's area (sensory integration)
Auditory cortex (hearing)

To answer this question Susan McConnell conducted an experiment: Neurons are born from the divisions of stem cells, the progenitors of all the tissues of the body. Stem cells can transform themselves into any kind of cell—blood, skin, heart, bone, or brain cell. McConnell took a stem cell before it morphed into a neuron, transplanted it into a developing brain, and watched it migrate. She found that its neighbors determined the fate of the soon-to-be neuron, the transplanted stem cell. The stem cell not only traveled alongside its neighboring cells, but took on the same function as its fellow travelers.

"The stem cell is still plastic," McConnell says, "still listening to signals from the outside. It gets instructions from the new neighbors and that sets its fate. It says, 'Yeah, sure, I can do that too.'"

In the second part of the experiment, instead of a stem cell, McConnell transplanted new brain cells waiting to undertake their migration outward from the neural tube into different parts of the developing brain. She found that the transplanted cells retained their original characteristics and did not become like their neighbors. "It appears that the brain follows the British plan. By the time a young neuron begins its migration, it has already received genetic instructions about what to become. In terms of that initial movement outward from the ventricular zone, genes determine a brain cell's fate."

But genetics is only part of the equation that governs human brain development. Simply put, there just aren't enough genes on the human chromosome to code for the placement of billions of neurons along with their trillions of connections. "If you had to program purely genetically every brain cell, without any sort of instructions from one nerve cell to another, you would have to have many more genes than are contained on human chromosomes," according to McConnell. What additional factors come into play?

More than Genetics In order to answer that question, envision an architect designing a six-story building (the cerebral cortex is made up of six layers). In this analogy the separate floors of the building result from the migration of the neurons to their predetermined positions. While the six layers of the cerebral cortex are taking shape, the "building" is gradually being "wired up." Neurons on each of the floors are making different connections. As a result, neurons in a particular area become specialized for mediating different functions such as vision (neurons located toward the back of the brain), touch, or hearing.

"Even though the six layers of the cortex look much alike, they differ just like the floors in our six-story building," observes Mary Beth Hatten. Different areas of the building—different areas of the cerebral cortex—are parceled out to specialize in mediating different functions such as vision or hearing or sensation or whatever."

For instance, projections from the eye link up with neurons in one of the brain's visual centers. Touch receptors in the skin eventually communicate with neurons located in the sensory cortex. Thus, while the brain's visual and touch areas each are composed of six layers and while each layer may look very much

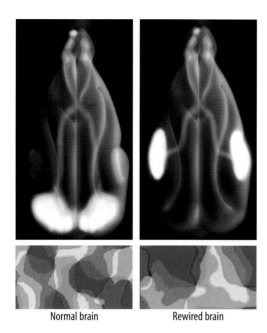

Normal brain   Rewired brain

Rewiring the Brain In an experiment to determine whether environmental cues can change the functional responses of specialized neural circuits in the brain, researchers altered the path of visual impulses from a newborn ferret's eyes. Instead of going to the visual cortex (far left), the signals were rerouted to the auditory cortex (near left). Neurons in the visual cortex normally respond according to the orientation of visual stimuli. For example, vertical edges, horizontal edges, or oblique edges trigger activity only in cortical cells that respond to that particular stimulus. Neurons in the auditory cortex normally decode sound.

Applying different colors to the neuronal activity related to the recognition of vertical (blue), horizontal (yellow), and oblique (red and green) edges, researchers are able to compare the cortical "maps" of the normal and rewired brains. The distinctive pinwheel pattern evident in both maps points to the ability of the auditory cortex to respond to visual stimuli. Such organizational maps occur in all mammals, including humans, but the patterns are distinct for each species.

alike under a microscope, each is involved in elaborating different functions and communicating with neurons in different parts of the brain.

But just how predetermined is this selective partitioning within the brain? Once the neurons have migrated and taken their place, do genes once again wrest back control from environmental influences? And if not, how much influence does the environment exert on the developing brain? Put differently, can nurture trump what nature intended? In order to answer that vexing question, Mriganka Sur, a neuroscientist at the Massachusetts Institute of Technology, turned his attention to the lowly ferret.

"We know that the brain has different areas that do different things, such as the visual cortex at the back of the brain or the auditory cortex in the middle of the brain," says Sur. "But what happens if we rewire the eyes so that visual inputs instead of going to the visual cortex actually go from the eyes to the auditory cortex that normally processes hearing?"

A Rewiring Experiment To find out, Sur rewired the brains of newborn ferrets. In Sur's altered ferrets the visual impulses went to the auditory cortex. "We reasoned that if we can make visual inputs from the eye go to the auditory cortex, which normally processes hearing, then the auditory cortex would become like a visual cortex. We predicted that the hearing cortex would come to have the same circuits and connections that mark the visual cortex."

By recording a series of images directly from the brain of his rewired ferrets, Sur showed a pattern similar to the pattern observed in the visual cortex. "The auditory cortex has been transformed by receiving visual signals, but not entirely. The characteristic pinwheel pattern is present, but it is less orderly, less graceful. As a result, the ferret's vision is closer to 20/60 rather than normal 20/20 vision."

Sur's experiment demonstrates not only the remarkable plasticity of the brain, but also the mutual interaction of nature and nurture. While the genes play the role of an architect who sets down the structural blueprint, the environment acts as a creative interior decorator who transforms that architectural design into the most marvelous edifice in the universe: the human brain.

But what happens when something goes wrong, when some deviation occurs from the original architectural plans? "The middle trimester of pregnancy, the period from the twelfth to twenty-fifth weeks, is when the fetus's brain is most vulnerable," according to Susan McConnell. "As the early cells destined to become brain cells start to divide, they can be killed by X rays or toxins. And such exposure can wipe out cells destined to form an entire part of the brain, resulting in severe mental retardation. Such babies were born after the mother's exposure to the atomic bombs dropped on Hiroshima and Nagasaki, or more recently, babies born in the aftermath of the mother's exposure to the nuclear reactor accident at Chernobyl."

Migration Errors Defects in neuronal migration can result in devastating conditions like childhood epilepsy, mental retardation, and fetal alcohol syndrome. Alcohol, when used by the mother during her pregnancy, interrupts migration by a "kind of random hit," according to Hatten.

"It's like a random act of terrorism directed against the developing brain. No one knows exactly what cells are developing in the brain on a given day. Maybe they are the cells that are somehow important in playing the piano or doing calculus. If the mother is drinking heavily, the alcohol will affect those cells like a terrorist's bomb. When the child grows up he or she may have problems with music or math. Or the child of a drinking mother could have other complicated difficulties along the fetal alcohol continuum that will be hard to understand and to help."

On the most basic level, the children born to alcohol- and drug-abusing mothers will have difficulties because the alcohol or drugs have affected the fetal brain and interfered either with the migration of neurons or with their "wiring." "Unless the brain is wired properly, it won't work," according to Carla Shatz, whose research focuses on how the immature brain wires itself up. She breaks the wiring process into two phases.

In the first phase, as we have noted, genes specify the directions for growing axons to take. The directions are so precise that most connections are accurately made with their target. Shatz compares this first phase to the establishment (in the pre-cell phone era) of telephone trunk lines linking two cities.

"You can imagine the first stage of brain wiring like connecting phones in New York and Boston. You want to make sure these connections are correct so that you don't wind up connecting New York and, say, Washington by mistake. The wiring is specified genetically, with the connections following very defined rules. 'After leaving the eye via the optic nerve, turn right at the optic chiasm, cross the chiasm, head toward the lateral geniculate nucleus, and grow into that nucleus. Don't grow into the medial geniculate nucleus because that's an auditory rather than a visual structure.'"

Next comes the second phase, which Shatz refers to as "address selection": "Now, with the wires and connections in place, imagine that you want to place a phone call from your home in Boston to a friend living in New York at Park Avenue and 47th Street. With a properly functioning phone system, making

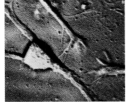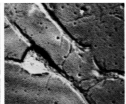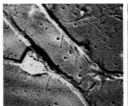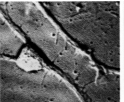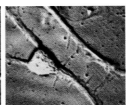

This series of spectrograph images shows axons and dendrites forming connections. Predetermined biochemical cues cause the neurons to grow toward one another, ultimately forming the link that will allow impulses to travel freely between them.

that call isn't a problem. You just pick up the phone and dial your friend's number and his phone, and his phone alone, will ring. But imagine a primitive phone system in which not only your friend's phone rings when you place your call, but so do a lot of other phones in New York City. Such a primitive phone system contains a lot of redundant connections that have to be eliminated. This is done in the brain by strengthening certain connections through repetitive use and eliminating redundant connections. It's as if each time you dial your friend's number, the specific connection between your phone and his is strengthened, while all other connections are weakened and, thanks to disuse, eventually eliminated."

Pruning Neuroscientists use the term *pruning* to describe this process of selecting appropriate connections and eliminating redundant ones. The process takes place throughout the brain during early development.

Carla Shatz learned about pruning through her research interest in human vision, specifically, how cells in the retina link up with brain cells in the visual cortex. The process takes place in the womb and involves electrical and chemical signaling.

Returning to the telephone analogy, it's as if fetal brain cells are pruning themselves by a process of "phoning home." For instance, about a million nerve cells in the eye connect up to 2 million nerve cells in a relay station from eye to brain known as the lateral geniculate nucleus. But many more connections are made than are actually necessary and many millions must be eliminated.

"Amazingly, this process of strengthening and pruning, this whole process of phoning, starts in utero during the first half of development," says Shatz. But, as we asked her during our interview, Who is placing the phone calls? It's dark in the womb and, besides, the light-sensitive cells in the eye haven't even been formed yet. So where are these phone calls coming from if they're not related to vision?

"What's happening is the ganglion cells in the eye are spontaneously 'autodialing' and connecting to target cells in the brain's visual pathway," Shatz told us. "They're sitting there in the dark spontaneously placing 'phone calls' into the brain practicing for vision."

Practice Drills   Nor is this "autodialing" limited to the visual system. Developing cells in other parts of the fetus are also carrying out practice drills in preparation for later experience. For instance, nerve cells in the ear are sending signals to the auditory parts of the brain even before the formation of the ear's sound-sensitive hair cells. Thus, before the fetus can hear, it's already practicing for hearing. And in the spinal cord spontaneous neural activity takes place that establishes the circuits needed for later movement.

"So the baby's brain is really a dynamic structure that's constantly changing in response to this phonelike process of running test patterns and remodeling connections among neurons," according to Shatz.

At the operational level neurons "talk" to each other throughout early development in the language of chemistry.

"The language that neurons use to communicate involves molecules secreted by one cell, 'Hello,' and a receptor on the surface of another cell that binds that signal, 'Yes, I'm hearing you,' " according to Susan McConnell. "When the signal binds to the receptor it sets off a whole cascade of events that alter the expression of genes within the receiving cell."

Among those neurons with similar firing patterns—signaling the brain together—the connections are strengthened. "Neurons that fire together wire together," as one neuroscientist puts it. This aphorism is a variant of another one about the brain that we will encounter at many points in this book, "Use it or lose it."

"The flip side of 'neurons that fire together wire together' would be 'out of sync, lose your link'," says McConnell. Cells that aren't firing together don't establish connections. Throughout development we have these two processes: the strengthening of some connections and the pruning away of others.

At birth the process of establishing and strengthening some connections and pruning out others continues. Only now the infant is no longer in the darkness of the womb; vision has become the guiding influence—the placer of phone calls.

What Does Baby See?   During the earliest weeks and months of the infant's life, normal visual experience is essential for the establishment of normal brain wiring. But what exactly does the baby see? Obviously one cannot answer this question by the usual methods of ascertaining visual acuity in an adult. Babies

Responding to the Human Face

Do infants show any preferences for looking at the human face? By one month of age an infant will look longer at a face or face-like stimulus, compared to any other object moving in the vicinity. Thus, mothers are probably correct when they claim a "special privilege" in regard to their infant's ability to recognize them. Indeed, it's likely that the infant's recognition of the mother occurs even earlier than one month of age but through non-visual perceptions such hearing, smell, and taste.

By four months of age, the infant will respond as an adult responds when shown a picture of a face turned upside down: Recognition of the inverted face takes longer, and identification errors frequently occur. While we can only guess at the infant's success in correctly identifying inverted faces, the increased time spent looking at these unfamiliar facial configurations suggests a recognition problem similar to that of adults.

By four months of age an infant will also more quickly recognize a face seen in the left half of space (the left visual field is mediated by the right hemisphere), compared to the same face shown in the

A baby's vision is the last sense to develop, thereby protecting the fragile infant brain from over-stimulation by damping down incoming visual stimuli. As a result, what a baby sees resembles a faded photograph (above). Nevertheless, by as early as two days of age, a baby can recognize his or her mother by sight alone.

right half of space. This is the same as the adult pattern—a right-hemisphere bias for processing faces.

Two months later, at six months of age, electrical recordings of the infant's brain will show not only a clear preference for looking at the human face, but now a decided preference for the mother's face compared to the face of a stranger. The electrical recordings will also show that the infant recognizes a familiar toy compared to a new one.

By seven months of age infants have progressed to the point that they can categorize and respond to basic facial expressions, such as "happy" or "sad." Siblings sometimes take advantage of this power of emotional recognition by "teasing" their infant brother or sister by feigning angry or fearful faces. While the younger infant may only stare back at the mimer, the seven-month-old will often scream or show other signs of distress.

don't answer questions, nor can they be relied on to look at eye charts. Instead, indirect measures of visual acuity must be used.

At McMaster University in Hamilton, Ontario, Dauphne Maurer investigates infant vision. "Whatever a baby is looking at is reflected off the center of his eye: We use these reflections to determine what babies are watching." As an application of this finding, Maurer placed objects in front of babies and observed whether the babies looked toward the objects. She found that as long as the light was sufficiently dim (babies are intensely sensitive to bright light) the infants stared at the object, sometimes for several minutes. It didn't seem to matter what she showed them (lines, circles, bull's-eyes, even golf balls), the babies continued their fascinated gaze. But since the babies also looked at black-and-white striped paper, Maurer decided to construct a striped paper eye chart.

At one week of age, the finest black-and-white stripes the infant can see are 30 times wider than those a normal adult can see: stripes about one-tenth of an inch observed from 1 foot away. By two months, the baby will stare at stripes half this thick. Over the next six months, vision progresses to that of an adult in need of eyeglasses. Not until age 6 does the baby's vision approach normal adult levels.

A Faded Photograph   "The baby sees the world like a faded photograph," according to McMaster University infant-vision researcher Terri Lewis. "If you take a color photograph and alter it so that it's all washed out, and if you then look at the faded photograph through a tube (because of the baby's narrow field of vision), that's what the world looks like to a baby."

Why do infants see so poorly? Because the center of the infant's retina, known as the fovea, contains immature cells; and since the foveal cells—the cones—respond to fine lines and colors, the infant with his immature cones is at a disadvantage in terms of acuity.

"This immature fovea, not the lens, is what limits a newborn baby's acuity," Maurer wrote in *The World of the Newborn*, a popular science account of the infant research, cowritten with her husband, Charles Maurer. "This immature fovea blurs lines and edges, and limits the perception of textures and details." In their book the Maurers provide a vivid depiction of what it would be like for adults with a similar lack of sensitivity to contrast.

"Textures would suddenly disappear: On a cloudy day, concrete would look as smooth as porcelain. Parts of many objects would meld into the areas surrounding them, and as a result objects would merge together. People's faces would lose their modeling. Sometimes whole objects would disappear into their surroundings. You would feel as if you were living in a drawing by Escher, a faded drawing with all the subtlety washed out."

"The fact that the baby doesn't see very well is, in a sense, very good for its immature brain," according to Lewis. "The immature brain can't handle excess stimulation. So the eye dampens down the visual stimulation to a level that the brain can handle."

Other aspects of the infant's visual world include a sluggish, inaccurate focusing system; an absence of stereoscopic, three-dimensional vision; extremely limited peripheral vision (Maurer says, "A young baby sees as though he is looking through a tunnel, or rather through two tunnels, since his eyes range independently over the world"); and an overall appreciation of brightness and, in consequence, extreme sensitivity to light.

Vision and Brain Development It's important to remember that the infant's vision is directly correlated with brain development. The visual cortex is especially immature in the infant. Cells aren't yet segregated by type or lined up into columns as in the adult brain, but are intermixed. Most of the cells lack a myelin sheath, the fatty substance that enhances rapid communication from cell to cell.

Since eye and brain develop in close harmony with each other, any interference with the passage of light from the eye to the brain results in serious disturbances. A cataract—a clouding or opacity of the lens of the eye—poses the most common threat to normal visual development. If the cataract isn't operated soon after birth—within a few weeks at the outer limits—blindness in the affected eye may result.

More than 30 years ago David Hubel and Torsten Wiesel provided the first hint of the importance of early visual experience. In a now classic experiment using kittens, they demonstrated that normal vision fails to develop if the kitten brain fails to receive visual information during a specific critical period early in the animal's life. Specifically, if a kitten's eye was sewn shut at birth and the sutures

Helping Babies with Cataracts

During the earliest weeks and months of an infant's life, normal visual experience is essential for the establishment of normal brain wiring. If a baby is born with a cataract, the affected eye does not transmit stimuli to the brain and thus the brain will not make the connections to the visual cortex. The unaffected eye then takes all the cortical circuitry used for vision.

If infant cataracts are not treated as soon as possible (within the first few weeks of life), blindness in that eye will be the likely result, even if the cataract is later removed. Thus it is crucial that the developing brain receive input from both eyes in order to wire the brain correctly for proper vision.

After the cataract is removed, the unaffected eye still holds an advantage over the weak eye, since it had a head start in forming the brain connections to the visual cortex. Thus, after the operation, the baby must wear a patch over the strong eye for several hours a day, in some cases for a period of several years. The weak eye can then play "catch-up" and improve as the neurons receive all visual stimulae and convert them to pathways between the eye and visual cortex.

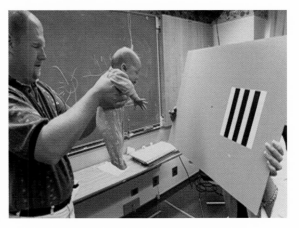

After having a cataract removed, the baby above was fitted with a contact lens in the affected eye. To test improvement in the vision of the now functioning but weaker eye, a doctor shows the child a series of cards with black lines that appear in different positions. With a patch over the stronger eye, the child must use the weaker eye to track the movements of the lines.

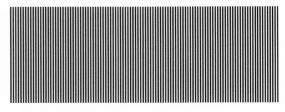

Using black vertical stripes on a white background, researchers have determined the width the stripes must be for a baby to see them from a particular distance. For example, at one week of age, the finest stripes an infant can perceive from four feet away are two-fifths of an inch wide (above, left);

this is 30 times wider than the stripes an adult with normal vision can see from the same distance (above, right). During the next six months, the baby's vision improves to that of an adult in need of glasses. Not until age six does the baby's vision approach normal adult levels.

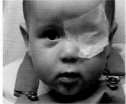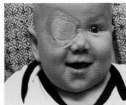

removed after eight weeks, the kitten's brain no longer responded to stimulation from that eye. But if the eye was reopened sooner, during the eight-week "critical window," the brain retained the ability to respond to light striking that eye.

In essence, Hubel and Wiesel demonstrated the existence at birth of circuits of cells in the kitten's brain that are ready to respond to visual stimulation from both eyes. But if one of the eyes is deprived of visual input during those critical first eight weeks of life, the neurons associated with the shut eye lose their connections with one another and the neurons associated with the open eye take over. According to David Hubel, "The closure of one eye produces effects largely as a result of competition. It is as though the neurons from the open eye had somehow taken advantage of their rivals from the closed eye by making more connections during the time of the 'critical window.' In the competition for space in the visual cortex, the neurons from the open eye had won."

Sensitive Periods On the basis of the work of Hubel and Wiesel some scientists speculated for a time that "critical periods" exist for such things as learning and the establishment of emotional attachment of a child to its parents. As we will discuss further, a more accurate description is "sensitive" period rather than "critical" when referring to most brain processes other than the early establishment of visual pathways within the infant's first few weeks of life.

According to Carla Shatz, "When an infant is born with a cataract on one eye, the eye loses its connections with the brain. Meanwhile, the other eye continues to establish and prune its connections and thereby takes over all of the cortical circuitry that's used for vision. That's why a cataract poses such a threat to the infant but not to an adult. If the infant's grandmother develops a cataract, there isn't any rush to get it operated on because all of the connections between Grandma's eye and brain were established many years ago and they remain stable over time. But when the infant has a cataract, the 'use it or lose it' principle comes into play. That's because the early developmental period is heavily involved in the selection and pruning of connections. And if one eye isn't used during that time because of a cataract, the other eye wins in the game of pruning. It takes over all of the cortical circuitry for vision that's ordinarily shared by both eyes."

With so much at stake in the affected eye (vision versus blindness), surgeons, understandably, elect to remove cataracts as soon as possible after birth, with best results obtained when the surgery takes place within the first two to four weeks. Time is of the essence because the longer the delay in removing the cataract, the greater the likelihood of permanent brain damage secondary to the loss of the early visual experience so essential for normal development. Missing visual experience from birth for just one or two months can result in permanent consequences for normal brain wiring and for later brain performance.

"The earlier your surgery, the sooner the brain cells in your visual cortex receive their normal stimulation and develop normally," according to infant-vision researcher Terri Lewis, who conducts extensive follow-up examinations on babies surgically treated for cataracts. If the cataract is diagnosed and operated on soon enough, the infant will have normal vision by one year of age. But in order to bring this about it's necessary to patch the "good" eye for several hours a day.

"If you have a cataract in one eye and surgically correct it," Lewis told us, "you still have an imbalance. The formerly deprived eye has to compete with the other stronger eye that has already established a lot of cortical connections. But you can correct for this imbalance if you patch the good eye. Patching gives the deprived eye a chance to exercise, a chance to work, a chance to gain some cortical connections."

Typically, a pediatrician or family doctor diagnoses a cataract by looking into the infant's eye with a specialized instrument, an ophthalmoscope. If the eye is normal, the doctor can get an unobstructed view of the area where the optic nerve leaves the eye and enters the brain. But if the lens of the eye is clouded, the doctor can't see the optic nerve.

Use It or Lose It Alex Levin, an eye surgeon at the Hospital for Sick Children in Toronto, specializes in the removal of cataracts in young infants. During our discussion he reminded us that vision requires not only eyes that can deliver perfectly clear images, but also a brain that's sufficiently developed to read that image. "Certainly the relationship of eye to brain is an intricate one. For instance, suppose I show you an ugly picture while you're asleep. You won't react to that picture at all, even if I hold your eyes open. Your failure to react isn't because you're blind while sleeping but because your brain isn't registering the image. The same thing is happening with a baby who has a cataract. Images critical for normal brain development just aren't getting there. As a result, the brain isn't provided the opportunity to go through its normal visual development."

When Birth Occurs Too Soon   Vision is not the only area where normal functioning requires the timely and accurate establishment of circuits within the brain. When a child is born prematurely, the process of "wiring" the brain is only partly completed. As a result, the establishment of a huge percentage of brain circuitry must take place within the nursery.

Thanks to the technological wonders of the modern intensive care unit, most premature babies survive. But a large proportion experience difficulties later in life involving paying attention or learning. In order to understand why, it's helpful to think of premature infants as fetuses who are forced to develop in extra-uterine settings at the time when their brains are growing at a more rapid pace than at any other time in their lives.

Prematurity exerts two main negative effects on the developing brain. First, the premature brain is more sensitive to events and processes that a full-term infant of this age would not normally encounter. This can range from indigestible nutrients (secondary to an underdeveloped digestive system) to brain hemorrhages. Second, premature birth interrupts the normal process of intrauterine brain development; growth factors provided to the premature infant's brain are missing, along with normally available intrauterine stimulation.

Included among the other difficulties experienced by premature babies are instabilities of temperature regulation resulting from a deficiency of body fat, impaired breathing as a result of immature lung tissue, blood pressure and pulse irregularities resulting from undeveloped self-regulatory processes, and feeding difficulties resulting from the combination of the infant's underdeveloped gastrointestinal tract and its weak sucking reflex.

Sudden alterations in blood pressure and pulse are particularly perilous. During episodes of apnea, or stopped breathing, blood pressure increases as does the pressure within the premature infant's brain. The accompanying rise in the level of carbon dioxide in the blood causes additional stress that too often leads to a rupture in the thin blood vessels serving the premature infant's brain. Intracranial bleeding is still a principal cause of death and disability among prematures.

Over the past two decades, technological advances in neonatal intensive care units (NICUs) have greatly improved the odds that a premature infant will survive. Today more than 95 percent of infants born after 28 weeks of gestation and more than 50 percent of infants born at 24 to 28 weeks survive. But the odds are less favorable for premature infants born between 22 and 24 weeks, when mortality remains high.

Nevertheless, while neonatologists have made great strides in reducing the incidence of neonatal death, they have been less successful when it comes to favorably influencing neurobehavioral issues like learning, reading, paying attention, and regulating emotions and behavior.

A New Approach Heidelise Als, a developmental psychologist working out of Boston's Brigham and Women's Hospital and Children's Hospital, is convinced that improvement in these "subtler parts of function" will only come about when NICUs mimic as closely as possible the conditions normally existing within the womb.

"The infant in the womb can rely on the accumulated effects of thousands of years of human evolution to help in the formation of a well-developed, healthy brain. After cell migration each nerve fiber becomes encased in its myelin sheath so that information can be transmitted efficiently. But in the immature brain, the myelin doesn't develop as well, and many of the connections aren't made. As a result, the premature baby doesn't have the buffers and filters that are available to a full-term baby. Sigmund Freud once referred to the womb as the baby's 'stimulus barrier,' which protects the brain from too much stimulation too early. The premature infant is deprived of that barrier before its brain is ready."

Absent the stimulus barrier, the premature isn't equipped to handle light that is transmitted from an immature retina to an equally immature brain, which because of its immaturity isn't prepared to process vision. The premature brain can process touch, its own bodily movements in the amniotic fluid, even taste, smell, and sound, but it's overwhelmed by light. Add to this the environment of most NICUs: high-pitched, unpredictable, sharp sounds; clutter; a general flurry of activity by the ICU personnel—in essence, too much too soon of the "buzzing,

Thanks to modern intensive care units, premature babies have a good chance of surviving. However, the harsh light and loud noises of a standard intensive care unit (right) can overwhelm the vulnerable brain of the premature baby. In an effort to accommodate the fragile state of the premature baby's incompletely developed brain, Heidelise Als of Boston's Brigham and Women's Hospital and Children's Hospital has developed a softer approach to caring for these infants (opposite).

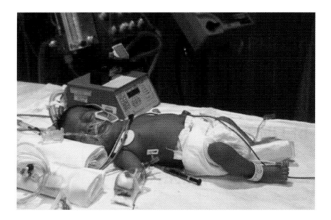

blooming mass of confusion" that William James described as typical of the mental life of infants. Actually, James's description applies most appropriately to premature, rather than full-term infants.

"Time is needed for the full intricacy of the baby's brain to form. That is one of the reasons premature birth is so dangerous," says Dr. Als. "The development of the brain must now happen in an environment drastically different from the dark, muffled, floating world inside the womb. Typically, premature infants are thrust naked under the bright lights of a newborn intensive care unit, their bodies invaded with tubes, overwhelmed with sensory information."

Mimicking the Womb   In order to help the infant born before his or her time, Als and her team at Brigham and Women's Hospital set out several years ago to create a "womblike" environment within the NICU. To envision what that would be like, consider the following:

Life within the womb is quiet, dark, and relatively silent—an appropriate environment for development of the premature infant's brain. Consider that while dwelling within the womb the fetus's most sensitive areas include around the eyes, within and around the mouth, on the palms of the hands and the soles of the feet, and at the genitalia. As expressions of this sensitivity, the baby floating within the amniotic fluid often clasps one hand to another, moves a hand to his face, or touches his buttocks or feet. She rolls and extends and stretches out, arching backward with the agility of a gymnast. And while performing these gymnastics the baby may suck on the ball of her thumb or her wrist. During all this activity, the infant's eyes are open and engaged in active eye movements, even in the murky darkness where little or no light shines (some scientists believe

The idea is to emulate the conditions of the womb as much as possible, with prolonged skin-to-skin contact, muffled sound, and muted lights. Early results of this work suggest that babies who have been given this extra care leave the hospital earlier and switch from tube feeding to bottle feeding more quickly. Moreover, months later these babies are developmentally more mature compared to premature babies who have received standard intensive care.

a tiny bit of light might reach the fetus if the mother's abdominal muscles are sufficiently thin).

By 18 weeks, sounds can be detected, especially within the range of the human voice. When the mother speaks the infant can hear her voice coming not only from outside the womb but also from inside it, since the mother's voice resonates within her body cavity. Over time the infant will learn to distinguish Mother's voice from the voice of another woman.

Creating a Special Environment   Translating all this into practical care methods, the premature infant should be shielded from the stress of sensory overload. Some freedom of movement is preferred to rigid swaddling. And the premature infant should regularly hear the comforting sound of Mother's voice. Ideally the NICU environment, like the womb, should be dimly lit and quiet, an oasis in which the baby's brain can mature without the overstimulation of too much, too soon. The baby should be held or supported to the limits of her tolerance (as a compensation for loss of the comfy confines of the womb). The parents, particularly the mother, should be with the infant and encouraged to simply 'Relax. Lie back. Breathe softly, enjoy being with your baby."

To provide this special environment, and thereby compensate for the vulnerabilities of the premature brain, Dr. Als and the staff at the NICU at Brigham and Women's Hospital have developed an innovative approach. The tiny infants, many weighing slightly over a pound, are swaddled in cotton flannel and placed in a quiet room with a blanket over the isolette to shield them from bright light. Natural bodily rhythms, rather than preset schedules, determine the "preemies' " care. For instance, the infants are permitted to decide when they want to eat rather than conforming to the rigid demands of the hospital-imposed feeding schedule. And rather than spending most of the day lying alone, they are cuddled and cared for by their parents, who hold the babies against their bodies and gently rock them when they need comforting. In order to create a quiet, calm atmosphere, the staff speaks softly, and radios or televisions are strictly forbidden. Parents are instructed to avoid engaging their infants in the kind of intense eye contact parents traditionally employ with their full-term infants. These special arrangements, evolving from the work of Als, aim at creating an atmosphere that supports the preterm infant in ways appropriate to the development of the brain.

"Premature infants are hypersensitive, easily stimulated, more vulnerable to noise and sudden movement," says Dr. Als. "The constant bombardment of stimuli in the typical NICU keeps the premature infant in a hypersensitive, hyper-responsive state, like that state experienced by an adult after too many hours of high-intensive work in an environment marked by continuous interference, interruption, distraction, and unpredictability. In such an environment, a person is perpetually at the mercy of whatever's happening at the moment."

So what is the best approach to mimicking the magically free and contented experience of the womb within the confines of a noisy, too brightly illuminated, often chaotic environment of the NICU?

"You try to establish a relationship approach to newborn intensive care. The baby is saying, 'See me for what I am and use what I'm trying to do to my own benefit.' So, first of all, it's necessary to observe the baby. Its behavior continuously provides you with a little window on its brain. For example, if you position the baby in a way that it can't manage, the baby will briefly stop breathing. You must learn to interpret the baby's behavior rather than fitting the baby to the NICU schedule."

Positive Results Dr. Als has documented positive results from a relationship-based approach that keeps mother and baby together as much as possible. The babies breathe on their own earlier and can switch more easily from tube feeding to bottle feeding. And when held by the mother, the premature infant's blood oxygen level remains higher and more stable. The infants also leave the hospital earlier. At two weeks and again at nine months, after what would have

Developmental Milestones to Twelve Months

Although the following behaviors are considered typical for these ages, every baby develops at his or her own pace.

By 3 Months

- Eyes follow you as you move across the room
- Smiles in response to your smile
- When lying on stomach, can raise head and shoulders
- Turns head and looks in direction of a sound
- Coos and babbles
- Ceases crying when you enter the room and recognizes familiar faces and objects
- Actively holds rattle and pulls at blanket and clothes

By 6 Months

- Plays with his/her own hands
- Begins to imitate sounds
- Babbling contains sounds like ma, mu, da, di, hi
- Bites and chews
- Grasps and plays with small objects
- Rolls from back to stomach
- Sits in high chair with back straight and head steady
- When lying on stomach, can lift chest and upper abdomen off table, supporting his/her weight with hands
- Adjusts position to see a toy

By 9 Months

- Sits on floor for several minutes
- Begins crawling
- Pulls self to standing position and stands holding onto furniture
- Grasps small objects between thumb and index finger
- Understands "no-no"
- Looks for a toy if he/she sees it being hidden

By 12 Months

- Walks holding onto furniture or with both hands held
- Can hold crayon and make a mark on paper
- Can follow rapidly moving toy with eyes
- Says two or more words, besides "dada" and "mama"
- Recognizes objects by name and understands the meaning of several words
- Understands simple verbal commands (for example, "Give it to me," "Show me your eyes")
- Imitates animal sounds
- Shows emotions such as anger, fear, affection, jealousy

The Brain and Temperament

In general, no predictions can be made before about three or four months of age. Whoever first said: "All babies are the same" (probably not a parent) was most likely referring to the limited behavioral repertoire of the newborn.

Bodily states determine responses in the newborn and young infant. Thus, the earliest stages of emotional life (to the extent that we can infer an emotional life in the absence of direct testimony) center on feeling hungry or sleepy, too cold or too hot, too tightly constricted or insufficiently confined. Responses to these alternative states occur along a narrow scale from crying when distressed to cooing when satisfied.

Some infants rarely cry ("a good baby," as the proud mother describes him); others, diagnosed as "colicky," may cry for six hours or more per day. But early crying patterns don't provide reliable indicators of later temperament. Indeed, predictions about temperament can't be made until three to four months of age. Perhaps this is the reason some cultures celebrate a child's first birthday three months after the actual day of birth. By this age, marked changes have occurred in just about every aspect of infant functioning, ranging from observable behavior to physiological measures such as brain electrical activity.

By four months of age infants differ markedly in their reactions to events around them, their general mood, their activity levels. One temperamental pattern known as behavioral inhibition (fearful, shy responses to people and events) is characteristic of about 20 percent of healthy four-month-olds. What's more, this pattern persists into childhood and even adulthood. In many cases inhibited patterns are associated with enhanced right frontal lobe EEG activity. This appears to be linked with a tendency to become distressed or depressed in the face of stressful experiences.

For instance, Richard Davidson, a psychiatrist at the University of Wisconsin in Madison, finds that 10-month-old babies who cry when separated from their mothers are more likely to show increased right-sided prefrontal activation while at rest compared to infants who did not cry when separated from the mother. Moreover, this appears to be an enduring pattern. Among adults, Davidson finds that differences in emotional reactions among subjects viewing films depicting positive or negative scenarios (a party versus an auto accident) depend on their resting prefrontal activation patterns. Those with more right-sided activation are more likely to experience distressing emotions when viewing a negative scenario.

But caution is in order when it comes to predicting future temperaments based solely on observations at infancy and early childhood. While early temperaments may provide some indication of likely adult patterns, they only serve as a predisposition that can be affected both positively and negatively by influences unique to each infant.

been their expected delivery, the babies are developmentally advanced compared to their peers cared for in standard NICUs.

As an illustration of the differences, Als played videotapes for us of the examination of two nine-month-olds. Matthew, born 14 weeks premature and cared for in a standard NICU, lags behind the performance of a normal term infant his own age (nine months). He can't sit up without his mother's support. Nor can he focus and maintain his attention on a toy for more than a few moments. When the toy is moved out of his reach, he looks confused and starts to whimper. Anthony is also nine months old but, in contrast to Matthew, is a "graduate" of Als's individualized care program. He sits up easily and is immediately captivated by a toy, even reaching for it when it's temporarily out of reach. "Notice that Anthony doesn't fall apart the way Matthew did," Als points out to us. "Developmentally, he's almost normal."

Individualized care is now standard in half a dozen hospitals around the country. While it's too soon to know whether these preemies will grow up to be completely normal, the signs so far are promising. But the real question is: Does a relationship-based approach to newborn intensive care result in a better brain?

Michael Rivkin, at Children's Hospital in Boston hopes to answer that question by means of an ongoing study comparing magnetic resonance images (MRIs) of the premature babies with their full-term counterparts. He's concentrating on the gray matter (the neurons arranged largely in the cerebral cortex on the surface of the brain) and the white matter (all of the connecting fibers extending from the neurons to their target destinations). According to Rivkin, "Based on current data, a difference exists in the density of fibers in the white matter as they course along to their ultimate destinations. The amount and thickness of the gray matter also seems greater in the premature infants receiving the special relationship-based treatment."

Better Control of Behavior Enhanced cerebral development helps the young infant to grow into a child able to exert greater control over his or her behavior. "Should I do this or not?" Premature infants, as they grow older, experience difficulty in making these kinds of decisions and obeying social rules in general.

Frank Duffy, a neurologist at Children's Hospital, spelled out some of the social difficulties experienced by preemies during childhood: "Using electro-encephalograms (EEGs) and other tests that also measure the brain's electrical activity, we've found that preemies cared for in the typical NICU setting later show different electrical activity from full-term infants over their right hemispheres and over the frontal lobes." (In later chapters we will take up in more detail the role of the frontal lobes. But to summarize quickly, the frontal lobes are concerned with understanding social and behavioral rules and using these rules to plan and anticipate the consequences of one's actions.)

"These differences in electrical activity over the right hemisphere and the frontal lobes result in failures to understand social rules. For instance, when in a restaurant, one of these children might say out loud, so that everyone hears and turns and looks, 'Who is that fat lady over there?' These kinds of mistakes are related to what people call nonverbal learning disabilities. They result from difficulties emanating from the frontal lobes and from the brain's right hemisphere. By five or six years of age preemies show electroencephalographic (EEG) differences in these areas."

Duffy's studies, along with those of other neuroscientists, suggest that the prefrontal cortex (the most anterior area of the frontal lobes) is greatly at risk in premature infants. First observable at about 23 to 24 weeks, the prefrontal cells continue to develop well into early adulthood. But in preemies the prefrontal myelinization pattern is thinner, culminating in difficulties affecting concentration, attention, focus, and in the observed deficiencies in following social rules of behavior.

The results appear more hopeful in preemies cared for in the specialized NICU at Brigham and Women's, however. Duffy has found that a premature infant from the relationship-based NICU has an EEG that looks more like that of a recently born full-term infant than an EEG recorded from a preemie cared for in a standard NICU. Enhancement of brain development coupled with normal mental functioning—that's the goal aimed for by Als, Duffy, Rivkin, and others.

Frank Duffy: "In the early days of neonatology, just having these babies survive was a miracle. Now it's more or less expected that these 'miracle babies' will survive. But we have to go a step farther than just survival. We have to help these babies to be comparable to full-term infants. That goal is now within our grasp."

From Infancy to Childhood

There is good reason for the claim that early childhood years are formative. Indeed, virtually every human capability first makes its appearance during childhood. The child's first attempts at understanding the world occur and rapidly reach a stunning degree of sophistication. The child turns from solitary pursuits and interests to cooperation with others within a social network. Language is acquired, and powerful communicative abilities rapidly develop. And the child's emotional responses begin to include sadness, joy, and the other major emotions that define our species.

A glance at the list of developmental milestones on page 31 attests to a striking uniformity. Of course, a few perfectly normal infants and children acquire these developmental skills at slightly earlier or later times. But the developmental sequence is remarkably consistent. Does this imply that the overall uniformity in infant and child development is genetically predetermined? On the contrary, because of the plasticity of the human brain, we exhibit a lifetime capacity for change and reorganization. Indeed, it's the brain's very plasticity that provides us with the potential to overcome adverse influences at any time during our lives. But as we will see, this lifetime plasticity renders us both adaptable and vulnerable.

Syllable from Sound

THE CHILD'S BRAIN

At birth the baby's brain looks like a one-to-four scale model of the adult brain. Yet, unlike the adult brain, neurons in the newborn's cortex have formed only a few connections among themselves. But that changes during the child's first year. The connections between neurons multiply as the neurons sprout dense thickets of dendrites, the bushy fibers that receive impulses from other neurons, and axons, the long single fibers that carry the outgoing messages from each neuron to others spread throughout the brain. And within the first few months the number of synapses, the junctions where one neuron relays its message to another, multiply exponentially. Taken together, these changes transform the brain into what looks for all the world like a magic forest of incredible beauty.

But even during this period of luxuriant growth, the process of pruning is also occurring. Many of the early connections start to disappear in different areas and in accordance with different rhythms, according to University of Minnesota developmental psychologist and neuroscientist Charles Nelson. "And, it is the child's experience in the world that determines which pathways will be strengthened and which will disappear. It's as if the child's brain were preparing itself for very specific challenges. Over the years, as the same tasks are repeated over and over again, the connections will grow even stronger."

A sculptor working on a piece of marble provides a useful analogy for the pruning of redundant connections. Starting with a large block of undifferentiated marble, the sculptor chips away at bits of the marble to reveal a form. In this analogy, experience plays the role of the sculptor and determines which of the redundant neuronal connections to chip away because they aren't being used and which to leave in place because they are.

The Acquisition of Language   Nowhere is this counterpoint of growth and elimination more dramatic than in the way a child acquires language. At birth infants the world over learn to speak at roughly the same pace, regardless of the language spoken around them. This holds true even though some languages seem to differ radically from other languages (e.g., Turkish and English). Infants can do this because all languages, despite their superficial differences, are actually remarkably similar to each other.

In all spoken languages meaning is conveyed by means of critical sounds, *phonemes*. Corresponding to the letters in most written languages, phonemes are few in number (no more than about 200 different distinct sounds across all of the world's languages and about 38 different phonemes in English). Moreover, infants are born with an innate capacity for discerning differences among all the sounds used in the world's languages.

Adults, in contrast, have difficulty perceiving differences between sounds frequently used to distinguish words in a foreign language. For instance, if you're an English speaker learning Spanish as an adult, you'll find it challenging to reliably distinguish the difference between *b* and *p* in spoken Spanish. And adult native speakers of Japanese find it hard to discriminate between American English *r* and *l*, as in the words *rake* and *lake*. Indeed, native Japanese speakers hear both sounds as *r* since there is no *l* in Japanese.

By one year of age infants lose this keenness for perceiving sound differences across languages. Pat Kuhl at the Department of Speech and Hearing Sciences at the University of Washington, in Seattle, compared six-month-old Japanese and American infants. She found that the Japanese infants respond to the *r-l* distinction as accurately as their American counterparts. By 12 months, Japanese infants had lost this ability, while the American infants at that same age had become more efficient at discriminating between the two sounds.

Perhaps at this point you're asking the same question that occurred to me when I first heard of Kuhl's work. How were such findings determined? After all, six-month-old infants don't talk and can't be reliably depended upon to respond to words or sounds spoken in any language.

In order to determine the infant's responses, Kuhl outfits them with caps containing 20 electrodes for recording brain electrical potentials. With the caps in place the babies watch as an assistant entertains them with toys in an effort to distract them and capture their attention. At intervals, different sounds are played in the background over a loudspeaker. Simultaneously, measurements are made of the brain's electrical pattern from each of the 20 electrodes.

Some of the sounds have nothing to do with language while others involve critical speech distinctions such as *r-l*. Kuhl is interested in whether or not the baby's brain records a change when a physical change in the sounds is made (*r* changing to *l*).

"We've found that the infant brain is very sensitive to linguistic experience," she says. "Tests on seven-month-olds show brain changes regardless of whether the sound switch involves an English contrast or, say, a Mandarin Chinese one."

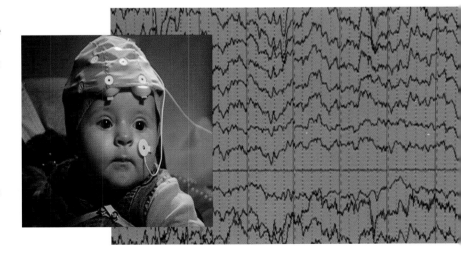

The world's people speak thousands of languages, and babies are born with the ability to master any of them. By monitoring infants' brain waves as they listen to different sounds (right), researchers have learned that babies up to the age of about seven months can distinguish between two sounds in, for example, Mandarin Chinese—a subtle difference their English-speaking parents cannot detect. By 11 months, however, unused neuronal connections have begun to be pruned away. As babies' brains become grounded in a single language, they cease to be "citizens of the world."

In her laboratory, Kuhl showed us brain recordings from seven- and eleven-month-old infants listening to subtle changes in sound involving Mandarin Chinese versus American contrasts. At about 350 milliseconds (thousandths of a second) the computer displayed a large shift in waveform regardless of whether the infant heard a contrast typical of the language spoken by his parents or a contrast from a language the infant had never been exposed to before.

Then Kuhl showed us the results from her testing of 30 eleven-month-old infants. When a change occurred in an English contrast, such as from *ba* to *wa*, the prominent deflection of the recording occurred at 350 milliseconds—an indication that the baby's brain was responding to the change. But when the same American babies listened to a contrast typical of Mandarin Chinese, the brain no longer showed a shift.

Says Kuhl, "The baby is no longer capable of processing differences in all the world's languages. At 11 months the baby's brain is very focused. The baby is listening through the filter that's beginning to be developed for their own language."

Kuhl's findings fly in the face of the traditional explanation of how infants learn their native language. Linguists once assumed that an infant's perception of sound began to narrow only after the infant learned some of the words in his native language. For instance, linguists assumed that children learn to discriminate between similar sounds by linking these sounds to word meaning—by learning,

Brain Cell Development from Birth to Age Two   A newborn baby's senses are flooded with the many sights, sounds, smells, tastes, and textures of the outside world. The baby's brain responds by growing more and more neuronal connections (right), strengthening some and losing others as the baby learns new skills and forms memories.

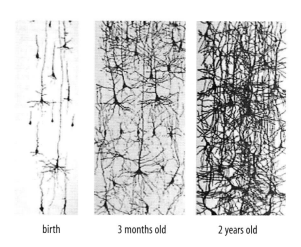

birth   3 months old   2 years old

for example, that "bit," "bite," and "beet" mean different things. According to this theory, if adults used the words interchangeably, children would treat the sounds of "bit," "bite," and "beet" as equals.

Incorrect Assumptions   This emphasis on word learning also provided the traditional but incorrect explanation of why certain sounds give rise to particular perceptual problems. A native Japanese infant, for instance, never encounters a word in which *r* and *l* phonetic alterations actually change the meaning of the word. This absence in the language of words with an *r-l* distinction was speculated to be responsible for the loss of the Japanese infant's phonetic perception of the difference between *r* and *l*. Linguists now know that these assumptions are incorrect.

According to Pat Kuhl, "Language input sculpts the brain to create a perceptual system that highlights the contrasts used in the language, while deemphasizing those that do not, and this happens *prior* to word learning. The change in phonetic perception thus assists word learning, rather than the reverse."

Kuhl also proposes that those sounds that are closest to those of the native language (prototypes) exert a "magnet effect" that works by attracting other sounds specific for that language. "What this means is that before infants understand or use language—before they utter or understand their first words—their perceptual systems strongly conform to the characteristics of the language spoken around them."

In a study comparing six-month-old American and Swedish infants, the perceptual magnet effect depended on the infants' previous language exposure. American infants showed the magnet effect only for American English. Swedish infants showed the opposite pattern: phonetic perception geared for Swedish rather than American English. "This makes sense," Kuhl says, "because babies must be able to learn whatever language their parents happen to speak. At birth babies are citizens of the world in terms of language. But by the time babies reach one year of age, they become language specialists. Babies born in Brooklyn start sounding Brooklynese."

But how does the infant become such an accomplished language specialist? Actually, the process is similar to what happens in the visual system. Neurons in the infant brain's visual cortex start out as generalists responding to gross aspects

The Brain's Language Pathways

Reading demands the sophisticated coordination of many structures of the brain. Indeed, it is one of the most complex cognitive acts our brain performs. Reading involves perceiving and identifying letters and words with parts of the brain—such as vision, hearing, judgment, and memory—that evolved for other purposes.

For nine of ten right-handed people and nearly two-thirds of left-handers, these language structures reside in the left hemisphere. When we hear and understand words, something called the auditory association area (just behind the auditory area proper) is active. But in order to read as well as to understand speech, we need a combination of visual and auditory processing. This occurs in the so-called receptive language area, which receives signals from both the auditory and visual association areas.

The receptive language area is often called Wernicke's area, after the German neurologist Carl Wernicke. In the late 1800s Wernicke's research laid the basis for much of our current understanding of how the brain encodes and decodes language.

When we speak or read aloud, Broca's area, the expressive language area located toward the center of the frontal lobe in the left hemisphere, comes into play. In the mid-nineteenth century, French anatomist Paul Broca discovered that lesions to this area in the left frontal lobe—but not to the corresponding area in the right frontal lobe—did not affect a person's ability to understand speech but did interfere with the ability to speak fluently and coherently. The discovery led Broca to make a now famous declaration: "Nous parlons avec l'hemisphere gauche!" ("We speak with the left hemisphere!")

Before we can utter any speech at all, however, the brain must first assemble appropriate words in Wernicke's area and then relay them to Broca's area for transshipment to the motor cortex that controls speech production. A bundle of nerve fibers called the arcuate fasciculus makes this relay possible. The fibers connect Wernicke's area to Broca's area via another brain structure known as the angular gyrus—key to linking letters with their associated sounds. The smooth functioning of all these parts is what makes bedtime story hour possible.

Reading aloud and being read to engage many of the same parts of the brain but in slightly different order. Though new research continues to reveal subtle nuances in the process, signals in the brain while reading tend to follow a set path. In the person reading aloud, the signal goes from the visual cortex to Wernicke's area. From there it travels via the fiber bundle called the arcuate fasciculus through the angular gyrus and on to Broca's area. After passing through Broca's area, the signal goes to the motor area, which initiates speaking aloud. For the person listening, the starting point is the auditory cortex. Comprehension occurs in Wernicke's area, and if the youngster wants to read along, signals go to Broca's area and then to the motor area.

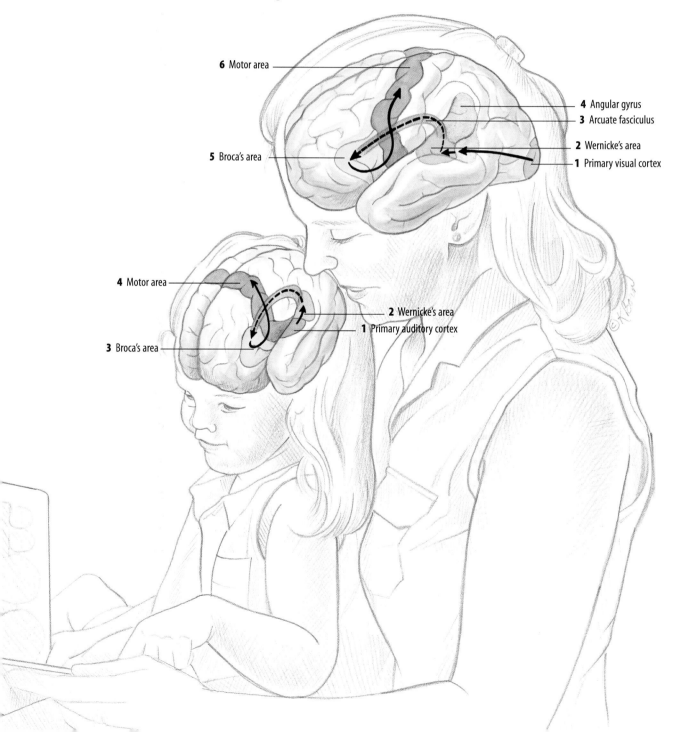

6 Motor area

4 Angular gyrus
3 Arcuate fasciculus

2 Wernicke's area
1 Primary visual cortex

5 Broca's area

4 Motor area

2 Wernicke's area
1 Primary auditory cortex

3 Broca's area

PET scans showing increased blood flow to active areas of the left hemisphere enable researchers to pinpoint some of the regions involved in different aspects of language processing. Hearing words activates the auditory cortex as well as Wernicke's area, near the junction of the parietal and temporal lobes. Silent reading engages the visual cortex, in the occipital lobe, and the association area for visual memory, which extends into the lower part of the temporal lobe. Speaking activates a region in the motor cortex as well as Broca's area, and thinking about words engages the limbic association area responsible for thought, learning, perception, and emotion.

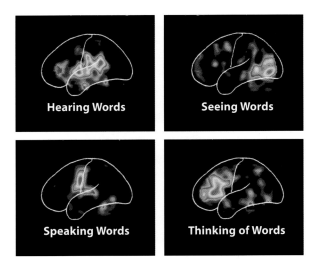

Hearing Words

Seeing Words

Speaking Words

Thinking of Words

of vision such as light and movement. But the neurons quickly become specialized and respond only to a single feature of the visual environment like orientation or direction. In the same way, newborns start out with their brains wired to hear a great variety of speech sounds. But over the first year the infant starts hearing only the sounds of the language spoken by the parents. It's as if the auditory system begins to sculpt itself so that eventually, at about the end of that critical first year, the infant perceives only the sounds of his or her native language. The connections among neurons associated with that language become stronger; the connections having to do with all other languages disappear.

Lip-reading provides another aid in furthering the infant's maturing mastery of its native language. In one test of this emerging ability a woman stands over an infant and mouths the sound *b* as in *bun*. At the same time a sound is played in the background that either matches *bun* or a different sound, such as *g* as in *gun*. In this test the infant looks longer at the woman's face when her silent lip movements match the vowel sound heard in the background.

The infant will put this new talent for lip-reading (a talent babies employ as early as four to six months) to good purpose throughout its life; visual information provides an auxiliary aid for adult speech perception, too. When adult listeners are given the opportunity to observe the face and lip movements of a speaker, their perception of what's said is greatly enhanced—the equivalent of a 20-decibel boost (quite vigorous) in the signal.

Guided by an already sophisticated knowledge of the sounds of language, 11-month-olds can mouth words such as *mama, dada*—to the delight of mothers and fathers throughout the world. And those enthusiastic parental reactions to their infant's first efforts provide reinforcement and encouragement for additional linguistic forays: new words, along with streams of nonsensical babbling that sound like the rhythms of more sophisticated linguistic productions.

Early Bilingual Exposure   Interestingly, the difficulty experienced by native Japanese speakers in making the *r-l* distinction isn't a problem if they have been exposed from birth to both English and Japanese. That's because infants who learn both languages early in life activate overlapping brain regions when processing the two languages, while those who learn their second language later in life activate two distinct regions of the brain for the two languages.

"The brain's processing of a primary language can later interfere with the second language," according to Pat Kuhl. "Acquiring new phonetic categories as an adult is difficult because the brain's mental maps for speech, formed on the basis of the primary language, are incompatible with those required for the new language—resulting in interference effects."

Infants exposed to two languages do especially well if each parent speaks one of the two languages rather than both parents speaking both languages. An American friend and his Russian wife provided confirmation of this general rule. She spoke to their eight-month-old son, Ivan, only in Russian, while my friend confined his communications to English (mostly because, as he put it, "my Russian is atrocious"). Ivan's comprehension of the two languages progressed smoothly except on those occasions when men visitors addressed the infant in Russian or women visitors spoke to him in English. At such times he turned fussy and restive. I was puzzled by this situation until I read Pat Kuhl's explanation of interference.

"It is easier to map two different sets of phonemes (one for each of the two languages) if there is some way to keep them perceptually separate. Males and females produce speech in different frequency ranges, and this could make it easier to maintain separation."

So Ivan's discomfiture isn't so inexplicable after all. A man speaking Russian and a woman speaking English was the exact opposite of what he was used to. As a result, he experienced an interference effect: a temporary difficulty separately perceiving and responding to the two languages.

Speaking Flawlessly   Within a month or so Ivan overcame the interference effect and responded to English and Russian no matter what the gender of the speaker. Most likely he will also grow up to speak both languages without an accent. In contrast, a child who learns a second language at puberty or later has much greater difficulty speaking without a noticeable accent, even with long-term language instruction. Indeed, a person's fluency in a second language depends upon how early in life he or she was first exposed to that language.

One study of language fluency concentrated on Chinese and Korean immigrants to the United States. At one end of the spectrum were immigrants who first heard English at age three; at the other end were some subjects who hadn't been exposed to English until age 39. All the volunteers evaluated recordings of spoken English for the purpose of detecting grammatical errors, such as incorrect word order or incorrect use of verb tense.

"We found a decline in correct responses as a function of the age at which people first arrived in the United States," according to Elissa Newport, a co-author of the study. "People who arrived before age five or so did as well as native speakers," but each group after that did systematically worse until the teenage years. At that point the response curve flattened out. "This is exactly what you would expect from a critical period."

Newport's reference to a "critical period" touches on one of the most controversial aspects of human brain development. As originally conceived, critical periods involved rigid cutoff periods that restrict learning to a specific developmental time frame. We've already commented in the last chapter on the most famous example of a critical period: the experiments carried out by David Hubel and Torsten Wiesel at Harvard in the 1960s. As you recall, these researchers showed that if a kitten's eye was closed soon after birth, the animal would be blind in that eye for life because the brain had been deprived of visual input during a "critical period" of brain development. Their research on cats led to the current practice of early cataract surgery on humans.

But while general agreement exists about critical periods for vision and perhaps some other kinds of fundamental brain development, most neuroscientists now favor the term *sensitive* periods when discussing most aspects of human development. And this distinction between *critical* and *sensitive* periods is important.

While it's true that exposure to certain experiences may be more effective in stimulating brain development if they come earlier rather than later, not all is lost when there are instances of delay. A variety of other factors can later make up for prior deprivations and ensure perfectly normal development.

"Sensitive periods define 'windows of opportunity' for learning," says Pat Kuhl. "During sensitive periods environmental stimulation is highly effective in producing developmental change. But the ability to learn is not equivalent over time." Nowhere is this more apparent than with language.

It's likely that Ivan's brain will physically incorporate English and Russian in different ways from an infant who is exposed to a second language a few years later. We know this because of the work of Helen Neville, a cognitive neuroscientist at the University of Oregon.

Left versus Right Neville compared brain activation patterns in Chinese and Spanish immigrants who began learning English at ages ranging from two to sixteen. In common with Elissa Newport, she noticed different responses when her subjects listened to sentences containing grammatical errors.

Among those who started learning English before age four, electrical measurements corresponding to error responses remained confined to the left side of the brain—the location of the language centers in most adults. But in the later learners—no English exposure until after age four—the activation originated in the right hemisphere.

"We are trying to understand how the brain's language systems develop," Neville explained during our interview. "Our subjects range from six-month-olds, who haven't yet started to understand or produce sentences, up to children about three years of age, who possess full-blown language capacities. We want to observe how the components of their language system develop."

Neville's use of the word *components* underscores an important point about language. Language involves more than just grammar. Also included are phonology (word sounds), semantics (word meanings), and syntax (word order). Neville and others have found that each of these components is handled differently by the brain and involves different language systems within the brain. In addition, the sensitive periods for the different components may vary widely. For instance,

Brain Hemispheres: Distributing the Duties For most people—whether they are left- or right-handed—language abilities are sited in the brain's left hemisphere. Interestingly, researchers have found that people who learn a second language after early childhood tend to process the second language in the right hemisphere. Generally, though, the right hemisphere handles less analytical, more intuitive abilities, such as flights of imagination, artistic or musical awareness, and psychological insight. In spite of this distribution, if one hemisphere is injured, the other can reorganize to take on some opposite-hemisphere tasks, depending on the age at injury.

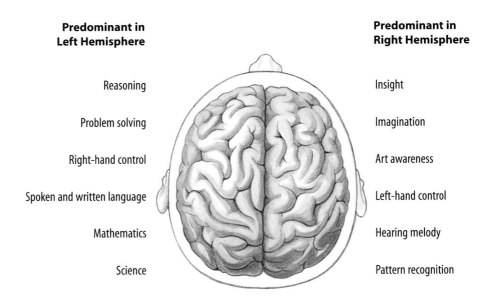

Predominant in Left Hemisphere		Predominant in Right Hemisphere
Reasoning		Insight
Problem solving		Imagination
Right-hand control		Art awareness
Spoken and written language		Left-hand control
Mathematics		Hearing melody
Science		Pattern recognition

while adults can learn the vocabulary of a new language without too much difficulty, most of us will likely have quite a lot of trouble handling the grammar and phonetics of the second language.

"We can learn a second language late in life without a great deal of difficulty in terms of vocabulary and the association of words with objects, people, and events—the semantic aspects of language. But the sounds of language prove very difficult to pick up later in life. That's why anyone who learns a second language after age 10 or so invariably speaks with an accent," according to Neville. And, as her own and Elissa Newport's research shows, later-life language learners are also less skilled in detecting grammatical errors.

"These different language systems—grammar, word meaning, word sound, et cetera—develop at different times. So there isn't just one critical or sensitive period for human language acquisition."

By the age of three, a child's brain is already using different circuits to process meaning and grammar. The pattern isn't like an adult pattern yet, but it's evolving in that direction. In addition, infants and young children don't seem to employ one hemisphere over the other for early language learning. A one-year-old child, for instance, uses both hemispheres when listening to new words. Six months later the balance is unevenly distributed, with preferential use of the left hemisphere for word processing. What's more, the timing of this increased language specialization in the left hemisphere corresponds to an increase in the number of connections (synapses) between neurons.

Learning Early, Learning Later   But what happens if a child's experience with language lags behind that of his peers? Does the left hemisphere still retain its edge in the processing of language? It depends. Neuroscientists are discovering that the age at which a child is first exposed to a language appears to affect not only facility in that language but also the brain's overall organization for that language. Rules of grammar prove especially vulnerable. For instance, among people who can speak both Chinese and English, it makes a great deal of difference at what age the speaker learned the second language (English).

When tested for their ability to detect grammatical anomaly errors (violations of phrase structure), early learners of English (delays of only one to three years before learning English) show the normally strong activation patterns observed in the left hemisphere of the majority of people. But among those children with longer delays before learning English (four to thirteen years), both hemispheres are activated.

A similar situation can be observed among the deaf. Neville measured increases in blood oxygenation levels measured by functional magnetic resonance imaging (fMRI) when hearing and deaf adults read English sentences. She observed that the brains of the deaf individuals responded like those of hearing people to nouns and other semantic aspects of language. But they responded differently to grammatical information. Specifically, they did not show the specialization of the anterior regions of the left hemisphere characteristic of people who have normal hearing and speaking.

Babies Babbling—in Sign Language

Most adults recognize the "ba-ba-ba" and other repetitive sounds that babies make as the beginning of their learning to talk. So familiar a phenomenon is this, in fact, that conventional scientific wisdom has long held that the onset of babbling is determined by the development of the anatomy of the infant vocal tract and the various other parts and structures of our brain and body that allow us to produce speech.

But Dr. Laura Ann Petitto, a professor in the Department of Education and the Department of Psychological and Brain Sciences at Dartmouth, has recently made some surprising findings. Petitto has been studying how young children who are learning spoken French and English or signed languages, including American Sign Language (ASL) and Langue des Signes Québécoise (LSQ) go about acquiring their various languages.

What she discovered is that deaf infants whose parents communicate in sign language babble with their hands in a way that is analogous to the vocal babbling of hearing infants. By systematically analyzing all the hand activity in her sample of hearing and deaf babies, she and her colleagues identified a discrete class of hand activity in deaf babies only that is virtually identical to descriptions of vocal babbling. Like vocal babbling, manual babbling occurs with a particular rhythmic frequency, in short phonetic units (syllables, as it were). Moreover, deaf babies even begin manual babbling at the same age (seven to eleven months) that hearing babies begin vocal babbling.

Petitto's finding confirmed the hypothesis that babbling represents a distinct and critical stage in the course of development of human language. By the same token, they refute the notion that the brain at birth is hardwired to acquire language via hearing and speech. Rather, Petitto argues, the human brain is hardwired to detect certain aspects of language itself—the natural rhythms and accents that identify language. If a baby's environment supplies those aspects of language, the baby will attempt to produce and to learn those patterns, regardless of whether they are produced by the tongue—or the hands.

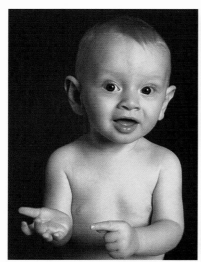

This six-month-old baby boy was caught in the act of producing rhythmic manual hand movements that possess the phonetic configurations common to manual babbling. Like hearing babies who seem to listen to themselves during solitary "crib speech," deaf babies look at their own hands when manually babbling.

Hands in Motion   American Sign Language, of course, uses visible signs rather than sound. Nonetheless ASL, like all languages, involves a complex grammar. But this grammar, in contrast to spoken language, makes extensive use of spatial location and hand shape and motion. We know this because of the pioneering work of Ursula Bellugi and her colleagues at the Salk Institute for Biological Studies in La Jolla, California. She showed that, in the brain, ASL has all the components and constraints (such as grammar and syntax) of spoken and heard language.

For instance, if a deaf person suffers a stroke in the "speech" areas of the left hemisphere, he or she suffers some impairment in sign production or comprehension. The situation is similar to the "aphasias" (impairments in language processing) observed after left-hemisphere damage in normally hearing people. In short, the capacity for signed and observed communication is represented predominantly in the left hemisphere in the same areas as those for spoken language. In addition, activation occurs in the areas in both hemispheres that are sensitive to motion detection. Or, to place the emphasis slightly differently, the language areas in the brains of both deaf and hearing individuals are specialized to handle symbols. In deaf individuals employing sign language, motion rather than speech that is spoken and heard provides the medium for this symbolic communication.

"The co-occurrence of location and motion information with ASL shapes the language systems of the brain, says Helen Neville. " Interestingly, hearing people who learned ASL (in order to teach deaf children, for instance) do not show any activation of the right hemisphere when reading or signing. This suggests, according to Neville, that a time-limited sensitive period may exist for the integration of the right hemisphere into the language system. If ASL is learned beyond this time, the normal left-hemisphere specialization for language will prevail.

When Language Centers Are Injured   If language takes its normal place in the left hemisphere, what happens when that hemisphere is compromised by injury or disease? As another feature of plasticity, the brain fights back.

At the Johns Hopkins Medical Center, in Baltimore, a team of neurosurgeons operate on children with a rare but devastating neurological illness known as

The Nicaraguan Town with Its Own Sign Language

Before the Sandinista revolution in 1979, Nicaragua had no means to educate the deaf. In addition, deafness had long been a stigmatized condition in that country, so deaf individuals tended to be hidden away in their homes. Few ever met other deaf people or became members of a deaf community.

When a child was born genetically deaf, his or her parents were usually not themselves deaf, but simply carriers of a recessive gene for the condition. To communicate with their families, these children used crude "homesigns"—gestures unique to each individual. Thus, when the government started new schools for the deaf, the children arrived without a common language.

But when American linguist Judy Kegl visited one of these schools in 1986, she discovered something amazing: Deaf children, never formally trained in sign language, were communicating just fine. The first group of children created a kind of pidgin sign language made up of their collective "homesigns." As younger children arrived, they added to it and were soon more fluent than the older children. In only a few years, the pidgin had exploded into a sophisticated and grammatically complex language. Now known as Idioma de Señas de Nicaragua (ISN, or Nicaraguan Sign Language), it is used by more than 800 deaf signers in Bluefields, Nicaragua, home of the Escuelita de Bluefields, the national school for the deaf. ISN has

Rasmussen's encephalitis. A virus-like illness, Rasmussen's destroys brain tissue on one-half of the brain, producing paralysis of the opposite side of the body, retardation, and seizures that cannot be controlled with medications.

Although the exact cause of Rasmussen's uncommon disease remains unknown, most experts consider it an autoimmune disease: The affected individual's body becomes allergic to its own brain tissue. And that allergy-like destructive response leads to an increasing number of seizures, sometimes hundreds in a 24-hour period that leave the young child—usually somewhere between five and thirteen years of age—completely unable to function in everyday life. He or she can't learn, loses the ability to speak, and eventually may even die in the throes of unceasing seizures. But neurosurgeons have devised a radical and, seemingly at first glance, unthinkable approach to Rasmussen's: surgical resection of the affected hemisphere —in essence, removal of one-half of the child's brain.

"It's better to have one-half of a brain that functions as well as it can than to have two halves but with one-half interfering with the function of the other," according to John Freeman, a neurologist at the Medical Center.

The interference results from the transfer of epileptic impulses across the corpus callosum (the band of fibers connecting the two hemispheres) from the diseased side to the normal side. And if that abnormal electrical activity—like static from a radio, as Freeman refers to it—originates in the left hemisphere, normal speech cannot develop. But with the removal of the diseased left hemisphere, speech originates from the right hemisphere.

a formal set of grammar rules, as well as written symbols, called "SignWriting" (opposite).

The deaf children in Nicaragua have afforded researchers a rare chance to study not only the processes of language acquisition but the invention of language itself. The experience has convinced Judy Kegl that both the ability to invent language—and a driving need to communicate—are built into the human brain.

Using Idioma de Señas de Nicaragua (ISN, or Nicaraguan Sign Language), a language invented by students at her school over the last 15 years, a nine-year-old deaf student communicates . On the chalkboard behind her are examples of "SignWriting," the written form of ISN.

"Language starts out bilaterally. It's on either side of the brain; but as our brain matures, the left side takes over. As a result, the right side largely forgets how to use language, forgets where things are stored."

The Uncultivated Right Hemisphere Freeman uses the analogy of a treasure hidden in a field of hay to illustrate: Normally, the left side of the brain is kept "mowed down," thanks to frequent language usage so that the treasure—*speech*—can be easily located. The right hemisphere, in contrast, isn't ordinarily used for language, isn't "mowed" or kept in shape and, as a result, turns into a large, uncultivated field. But if the left hemisphere is removed, the treasure can be found lurking within the remaining right hemisphere.

If you want to talk," Freeman says, "but can't do so out of your left hemisphere, then it's logical to try to do it out of your right hemisphere. And so you begin to search for the words in the right hemisphere. You beat down pathways through this field of hay and find where the treasures of language are hidden."

Of the several language "treasures" concealed in the "hay field" of the right hemisphere, not all are discovered simultaneously. Receptive language—the ability to understand what other people are saying—returns first. As a result, children who undergo left hemispherectomies (removal of the hemisphere) recover the ability to understand spoken speech as well as before surgery; indeed, they're eventually understanding speech as well as peers of the same age who haven't undergone surgery. But their ability to speak lags considerably behind.

Variations in Timing "Different aspects of language come back at different times," according to Johns Hopkins neuroscientist Dana Boatman, who tests the children after their hemispherectomies as part of her study of how children learn to understand spoken language. "We begin testing these children within the first month after surgery. At that point they're virtually mute, and their speech understanding abilities are very poor. But within six months after the operation, most of these children are back to their presurgery levels in terms of single-word comprehension. And they continue to progress, suggesting that the right hemisphere seems to do as good a job as the left hemisphere when it comes to understanding spoken language. Thus, the ability to understand spoken speech is not just in the left hemisphere but is represented in both hemispheres."

During our meeting, Dana Boatman spoke about Michael Rehbein, who underwent his first testing one week after his surgery. At that point Michael, although mute, could understand simple words. Six months later he could understand spoken speech at the same level as before his surgery. His speaking ability, however, remained slow and undeveloped. A year later, that too showed improvement.

"In every year since then his speaking ability has increased. So it seems that he is continuing to recover his language functions at different rates. For instance, Michael's recovery of speech was closely linked with the return of strength in his right arm and leg."

Why would the return of speech be correlated with improvements in motor control in the right arm and leg? Because speech is also a motor function: We employ the movements of our lips and tongue in order to form words.

"The motor activity of producing speech is closely tied up with the other motor functions involving the arm and the leg," explained Boatman. "That's the reason the recovery patterns occur over the same time course. As another possibility, the right hemisphere may be specialized for understanding speech, while the left hemisphere is more closely linked with producing speech."

Studies with fMRI suggest what's going on in Michael's brain as he recovers speech after his surgery. When Michael is listening to speech, activation occurs in the intact right hemisphere of his brain. Furthermore, the activation takes place in areas of the right hemisphere that correspond to the speech centers that were located in his now missing left hemisphere. "Our preliminary findings suggest that comparable areas in the right hemisphere are recruited for the recovery of speech understanding after the removal of the left hemisphere," Boatman told us.

In essence, the right hemisphere can serve early in life as a backup system, capable of taking on the language duties ordinarily carried out by the left hemisphere. But there is a time limit to the brain's plasticity for language development by the right hemisphere. So far, no child older than 15 has been operated on.

"Language understanding is present soon after the surgery," Freeman explained, "but in order for the child to find and produce spoken words, he or she has to search and find pathways through those hayfields in the right hemisphere. With younger children the hay hasn't grown so high, so they can make pathways more easily. Older children of 13 or 14 are able to do it too, but it takes much more time."

Declining Plasticity   To summarize, brain plasticity varies inversely with age: The younger the child, the greater the plasticity and the greater the potential for overcoming setbacks—even something as drastic as the removal of a whole hemisphere. Why should brain plasticity decrease with age?

Loss of plasticity is the price we pay to reap the benefits of specialization. Think of the brain as a structure composed of vast millions, even billions, of interconnected neurons. As with any similarly complicated structure, great benefits accrue from specialization.

For instance, the construction of your house or apartment depended on the contribution of specialists with varied interests and abilities. Architects and builders, electricians and plumbers, carpenters and landscapers—these are just some of the specialists involved. Now imagine what your house or apartment would be like if built by generalists instead of specialists. Under such an arrangement, all workers would know a little bit about each of the many specialized crafts, but no one would possess real expertise in any one of them. Most likely, you wouldn't want to live in a home built that way. The same principle applies to the brain: Optimal efficiency and performance results when, with increasing development, areas are parceled into specialized circuits. Fortunately, the partitioning of the brain into specialized zones is preceded by a long period of plasticity that, for some activities such as learning, continues over a lifetime.

Learning to Read   So far we have concentrated on only one aspect of language: the spoken (or signed). But in advanced cultures like our own, written language is equally important. And while speech will develop normally in the absence of any specific instruction, reading is an acquired skill.

Few of us can remember more than the sketchiest details of the long, arduous process we undertook while learning to read. We don't remember since for most of us (85 to 95 percent of the population) reading came fairly easily. And it was easy because most of us met certain basic requirements.

First, in order to learn to read, we had to know the letters of the alphabet and the meaning of a certain number of words. Next, we had to develop an awareness that the letters on the page represent the sounds of the spoken word. For instance, to read the word *dog* we had to parse (segment) the word into its separate phonemes (its 3 of the 38 distinctive sounds that make up spoken and written words in English). Once the word was recognized in its phonological form (a capability called "phonological awareness") we could identify and understand it. (For most children awareness of the phonological structure of words develops between four and six years of age—the age span when most of us learned to read.)

But reading doesn't come easily to everyone. Developmental dyslexia, as it's referred to by speech and language specialists, is formally defined as an unexpected reading failure that cannot be explained by low intelligence quotient (IQ) or environmental circumstances such as faulty teaching methods or a deprived upbringing. Somewhere between 5 percent and 15 percent of the general population suffer from the disorder, which tends to run in families and affects boys more than girls, strongly suggesting a genetic basis.

The Challenge of Dyslexia   For dyslexics the experience of learning to read often involves vivid and painfully humiliating memories of rejection associated with having to repeat early grades in school. Even the more successful of dyslexic children reads slowly and oftentimes inaccurately.

The first description of dyslexia dates back to 1896. In that year, W. Pringle Morgan, a physician in Sussex, England, described a 14-year-old patient in these words: "Percy F has always been a bright and intelligent boy, quick at games, and

A Child's Determination—and the Brain's Plasticity

Seven years after surgery to remove his left hemisphere, 14-year-old Michael Rehbein was spending summer Saturdays racing mini stock cars at the Tri-County Speedway near his home in upstate New York. If you asked him his favorite subject in school, the teenager would say, "I love math." The hardest thing, he'd tell you, is speaking. Although Michael's right hemisphere performs as well as a normal left hemisphere when it comes to understanding other people's speech, it is less adept at generating speech.

The decision to have Michael undergo surgery to remove the hemisphere normally responsible for speech was not an easy one. He was an athletic seven-year-old, climbing trees and playing ball. But he began suffering uncontrollable seizures—60 or 70 on a good day, as many as 300 or 400 on a bad day. Doctors said the only solution was to remove the site of the seizures, which, in Michael's case was the left hemisphere.

Recovery was painfully slow. "He had to learn to walk and talk," his mother recalled. "It was like he was an infant again." In fact, it was much harder. With his left hemisphere gone, the right side of his body was paralyzed. As his right hemisphere reorganized to take on unfamiliar tasks, he regained the use of his right leg. Learning to talk was even harder.

Six months after the surgery, he could make only one- or two-word utterances. But Michael was determined. One teacher who worked with him remembers the first full sentence Michael said to her: "I love you with all my heart."

While his doctors continue to monitor Michael's progress and to study how his right hemisphere has adapted, the teenager himself is getting on with life. "He's the same boy inside that he was before the surgery," says his dad. "He hasn't lost his drive."

Michael in post-operative therapy and (above, dressed for mini stock car racing) considerably further along in his recovery.

Michael's right hemisphere (near left) has reorganized to take over the language functions normally carried out by corresponding areas in the left hemisphere of an intact brain (far left). However, the right hemisphere is not as efficient as the left, and more areas of the brain are recruited to process speech.

When we engage in the act of reading, the brain must enlist a variety of functions, including attention, vision, and memory. In the images at right, reading requires, first, that we focus on the task at hand; paying attention activates the frontal lobes and, in midbrain, the thalamus. Next we must visually take in the words; this engages the visual cortex in the occipital lobe. To make sense of the words we see, the information must then go to an association area where we connect the symbols on the page to a word's meaning. In ways researchers still don't understand, the dyslexic brain makes a misstep along this path.

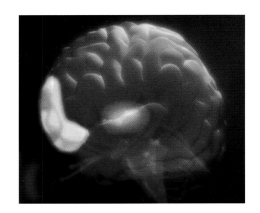

in no way inferior to others of his age. His great difficulty has been—and is now—his inability to learn to read."

What was the explanation for Percy's problem? Early theories placed the blame on the visual system. Visual defects were thought to explain why some dyslexics occasionally reverse letters when writing. Actually, letter reversal is rare and has nothing to do with alterations in vision. Nevertheless, eye-training exercises remained a favorite treatment for many years. Ophthalmologists treated dyslexic children by using cue cards with enlarged letters and words. But the method failed; most dyslexics have normal visual acuity. But they do experience difficulties binding words together into their constituent sounds (phonemes), the phonological awareness referred to above.

In dyslexia "phonological awareness" is impaired in brain processing. Thus the word *dog* isn't correctly decoded into its phonological form. As a result the child doesn't recognize the word *dog* on the page. What's more, this defect is strictly limited to reading; general knowledge is unaffected and may even be superior. The result is a puzzling contradiction between the child's general intelligence (often superior) and his grossly deficient reading ability. Thus the dyslexic child may not be able to read the word *dog* but when asked questions about a dog, he may launch into a vivid description of the virtues of his terrier, Toddy, followed by a discourse on the personalities and habits of various other dogs he has come into contact with. Such performances are less puzzling and contradictory when viewed as simply one reflection of a specific problem in converting words on a page into their constituent sounds.

Needed: A Good Memory Phonological awareness is only one component of successful reading, however. In addition, the beginning reader must possess a

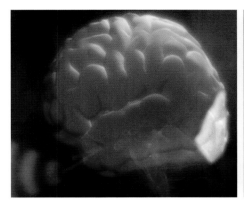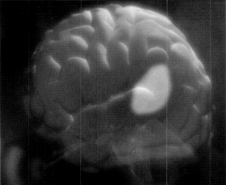

good memory. Short-term memory is needed to temporarily store the words, keeping them "online" so that sounds associated with letters found at the beginning of the word can be combined with letter sounds encountered at the end of the word. At a later point in reading development, the reader has to be able to extract meaning from the words based on earlier reading experiences.

"Reading is a very complex skill," says Guinevere Eden, of the Institute for Cognitive and Computational Sciences at Georgetown University, in Washington, D.C. "In addition to linking letters to sounds, reading involves remembering the visual image of a word and memorizing its meaning. The beginning reader must also learn to sound out letters and remember those sounds as she decodes the other sounds that go with the word. Finally, as she reads the entire sentence she must remember every word so that when she reaches the end of the sentence she can put everything together."

Most of us accomplished these complicated processes because we had no problem with the initial challenge: enhancing our skills in breaking down the words on the page into their constituent sounds. Over the years, reading specialists have identified several measures that can be applied to kindergarten children and accurately predict future reading and spelling skills in primary school. For example, the dyslexic child experiences difficulties in tests of phonological awareness that involve the isolation and manipulation of word sounds. Challenges such as "Tell me which of the following words doesn't rhyme with the others: hat, cat, dog, mat" prove difficult.

Another predictive test measures the child's ability to repeat back a word while omitting the first or last sound of that word. For example, *bat* without the first sound becomes *at*. Dyslexics have great difficulty carrying out such seemingly simple maneuvers. In a subtler example, the dyslexic child when asked to omit the first sound in the word *bone* says *one* as in *one, two, three*, et cetera, rather than the correct sound that rhymes with *own* as in "I own my house."

Predicting Future Skills While many theories have purported to explain the disorder—ranging from disturbances in visual or auditory processing, to defects in short-term memory, to timing errors in writing, spelling, and speaking—no single explanation is entirely satisfactory. For instance, while some dyslexics have trouble with short-term memory and can't hold words in temporary storage, others do just fine on this task and yet remain severely dyslexic.

"When you look at a group of individuals with dyslexia, you find that they all seem to have slightly different 'flavors' of dyslexia," says Eden. "Some of them are especially impaired in dealing with the sound structure of language—phonemic awareness, as we refer to it. Others have difficulty with their short-term memory and as a result can't deal with lists of words. Others can decode a word and can correctly identify how the word sounds, but they do so very, very slowly. And, finally, some dyslexics have all of these problems. In short, the under-lying basis of dyslexia may be different from one individual to the next."

Not surprisingly, such discrepancies among dyslexics have led to intensive efforts to discover the brain-processing difficulties responsible for the disorder.

The first correlation of reading with a specific brain area dates back roughly a hundred years, when a French neurologist, Joseph Jules Dejerine, encountered a patient who had lost the ability to read as a consequence of a stroke. He could speak perfectly normally and understand what others said to him, but he could no longer read even the simplest communication. At autopsy, Dejerine discovered a scar toward the back of the left hemisphere under an area called the angular gyrus. Contemporary imaging studies confirm the importance of the angular gyrus, which normally is activated when a person is reading.

Using Image Techniques Additional areas important for reading highlighted by imaging techniques are found along the length of the Sylvian fissure—the horizontal cleft along the outer surface of the brain that gives it the appearance of a wrinkled boxing glove. Finally, reading-associated activation occurs in areas involved in vision and the visual detection of motion. Variations from normal in the activation patterns from one or more of these areas often turn up on brain scans of dyslexics.

Since functional neuroimaging using fMRI is currently providing the most specific localizing information about the systems affected in developmental dyslexia, let's take a moment here to describe how the studies are actually carried out.

After placement within the scanner, a child considered at risk for dyslexia is asked to look at a small monitor and perform a series of manipulation tasks that serve as reliable indicators of dyslexia. The tasks require the child to work with phonemes (sounds making up a word). For instance, he or she may be asked to make judgments about words that rhyme (the *bat, cat, dog, mat* sequence). Next he looks at "pseudowords" (strings of meaningless letters that only look like words) and selects rhyming pairs *lete* and *jeat* but not *mobe* and *haib*. Or she may try reading aloud a series of words on a monitor and then repeat the words after deleting the first letter (*cat* becomes *at*).

She then hears rather than reads words and, once again, repeats the words without the first letter. Finally, the child watches dots moving across the monitor and identifies their direction of movement. While the child performs each of these tasks, the brain scanner registers her brain activity. The goal is to identify those brain areas that become activated when the child attempts the various tasks.

"We know that the ability to manipulate word sounds is very important in learning to read," says Eden. "So we are trying to identify where that skill resides in the brains of normal readers in comparison to dyslexics. So far, the imaging studies show a decrease in activity from both sides of the brain in the posterior temporal lobes and the inferior aspect of the parietal lobes."

A Surprising Finding   Overall, the fMRI findings indicate a neurological basis for dyslexia that isn't confined to areas traditionally considered specific for language. The moving-dot findings are particularly intriguing. Dyslexics do poorly in the seemingly straightforward job of estimating the speed of movement of some dots in comparison to other dots on the monitor.

On the basis of the fMRI findings, it comes as no surprise that dyslexics have trouble detecting differences in word sounds, or knowing how and where words are broken down, or identifying the sounds making up a word. But other findings, such as the difficulty estimating the speed of movement of the dots, are less easily understood. Nor are such difficulties confined to laboratory experimentation. For instance, dyslexic children often claim that the words on the page "look like they're walking." Does this strange claim result from a slowdown in visual processing? Guinevere Eden explains:

"Reading and writing make heavy demands on the visual system. For this reason, many kinds of reading errors can result from impaired visual processing, such as the difficulty with moving dots. A visual scanning error may cause the child to skip a line of words. Or the child may have difficulty integrating the look of a word on the page so that the child reads *dog* as *god*."

As we listened to Guinevere Eden speak about the dyslexic's visual difficulties, we thought back to Percy F. Based on their experience with early dyslexics such as Percy, authorities in the late nineteenth and early twentieth centuries emphasized—indeed overemphasized—the importance of the visual system. And while they have turned out to be wrong on the whole, they weren't entirely wrong. The findings on scanning errors mentioned by Guinevere Eden support the view that the visual system isn't entirely normal in dyslexics. The mistake of the early thinkers about dyslexia was one of disproportion. They overemphasized a minor component of the disorder while overlooking the main culprit: the child's problem breaking words and letters down into their component sounds.

As we will see at other points in our exploration into the secrets of the brain, overemphasis and underemphasis frequently create problems. For some reason, perhaps as part of our evolutionary heritage, we're attracted to straightforward explanations that involve small numbers of elements (an explanation involving only one element is most attractive of all!). For instance, many neuroscientists would like to settle on a single neurotransmitter or a specific lobe as the "explanation" for, say, schizophrenia. But the brain rarely complies with our need for simplistic explanations, even while it's true that the brain often employs a marvelous simplicity. As in dyslexia, most of the brain's functions in both health and disease involve many components working together. We've learned this from scanning results such as the fMRI findings in dyslexia. It has begun to be evident that dyslexia can involve functional disturbances in widespread areas throughout the brain.

Developmental Milestones to 36 Months

Although the following behaviors are considered typical for these ages, every child develops at his or her own pace.

By 15 Months

- Walks without help
- Can stand up without support
- Drinks from cup without help
- Begins using spoon but needs help
- Can place a round object in a round hole
- Shows intense interest in pictures
- Says four to six words, including names
- Asks for objects by pointing
- Kisses and hugs you, has temper tantrums
- Imitates activities such as sweeping and folding clothes
- Tolerates some separation from you

By 18 Months

- Walks up stairs with one hand held
- Throws ball overhand without falling
- Can jump in place
- Pulls and pushes toys
- Eats well with spoon
- Turns pages in book, two or three at a time
- Scribbles spontaneously
- Says 10 or more words
- Knows the function of common household objects (telephone, brush, spoon)
- Takes off shoes, socks, gloves, and unzips

By 24 Months

- Runs fairly well, kicks ball forward without losing balance
- Goes up and down stairs alone with two feet on each step
- Turns pages of book one at a time
- Turns doorknob, unscrews lid
- Has vocabulary of about 200 words
- Uses simple phrases
- Refers to self by first name
- Verbalizes need for toileting, food, drink
- Pulls people to show them something

By 36 Months

- Has daytime bowel and bladder control and may have nighttime control
- Rides tricycle
- Goes up and down stairs using alternating feet
- Draws simple designs like circle and cross
- Constantly asks questions
- Uses complete sentences of three to four words
- Recognizes and identifies most common objects and pictures
- Completes puzzle with 10 to 20 pieces
- Understands concepts of "mine" and "his/hers"
- Can take turns in games

Approaches to Treatment Armed with the scanning results, speech and language therapists are now in a better position to decide about treatment. Since the fMRI findings in two dyslexics may point to different problem areas, no single "magic bullet" works for both of them. Rather than one treatment, a menu of treatment approaches may be more appropriate.

One approach, the Lindamood Phoneme Sequencing Program (LiPS) aims at enhancing the dyslexic's awareness of the speech sounds that make up words and how those sounds are connected with the words. The goal, according to a brochure describing the process, is to help the dyslexic child to "become self-correcting in reading, spelling, and speech." In a nutshell, LiPS aims at improving the dyslexic's ability to articulate words—literally sound them out. The child is encouraged to become aware of the feeling of the word when it's said. In this phonological approach, input through the eye and ear isn't enough—the mouth must be engaged as well.

A second approach, the Seeing Stars program, emphasizes visual imagery. Most of us read by sight alone and rarely sound out words unless they're somehow strange or unfamiliar. But for some reason, the visual memory of many dyslexics isn't functioning well enough to grasp and retain the word. The Seeing Stars program attempts to correct this by enhancing the dyslexic's ability to visualize letters and words. For example, the child looks at a word for just a brief moment and is then asked to spell the word in the air with his finger prior to attempting to say the word. If this doesn't help, the child tries spelling the word in sand or on the tutor's arm or back. The goal is to encode the word into the child's brain by enlisting the help of the other senses.

"Our aim is to tailor the treatment according to the weaknesses," according to Eden. "And it's not just reading that's impaired. Fluency, working memory, phonemic awareness—any one of these may be at fault. You should take as your target which area or areas are the weakest. There's not one magic bullet or one type of intervention that will work for everyone."

In the near future it should be possible to correlate treatment responses using such approaches as LiPS and Seeing Stars with fMRI findings. As the child responds to intervention, the fMRI profile should change, thus enabling teachers and neuroscientists to construct a dynamic portrait of the brain changes accompanying improvement in a child's reading ability. So far, such correlations between reading improvement in and changes on fMRI are only suggestive and await the conclusion of ongoing long-term studies.

According to Guinevere Eden, "When a dyslexic has learned to overcome some of his reading problems, can we see consistent and reproducible differences in the fMRI? No one has done that yet, but researchers are working on it. The interesting question is: When you look at the fMRI of a dyslexic who has learned to read almost normally, does it look like the fMRI of a normally reading child, or has the child developed an alternative pathway? It will be interesting to find the answer to that question."

The Hyperlexic Reader   One path toward understanding the brain of both normal and dyslexic readers may come from the study of children like Alex Rosen. One evening, after dinner, Eileen Rosen began reading a fairy story to her two-year-old son, Alex. Within a few moments she discovered something very unusual about Alex:

"The fairy tale was very complicated, the kind you find in more difficult children's books. As I read the story I ran my finger along the lines. At one point, I accidentally skipped one of the lines. Immediately Alex took my hand and pulled my finger back to the correct words."

Eileen was puzzled. Alex hadn't yet spoken his first word and she reasoned, "How could he possibly be reading prior to learning to speak?" To find out, she decided to perform a simple test of her own devising.

"I purposely skipped several words on the page, and every time he pulled my finger back to the skipped words. In essence, he was communicating to me that, although at that time he couldn't talk at all, he could read well enough to absorb new information."

As an additional test of Alex's reading ability, Eileen wrote on a piece of paper the words *transportation, jewelry, furniture,* and other categories. When handed a piece of paper with the word *train* written on it Alex immediately placed it on *transportation*. "He was obviously processing the information correctly, but he couldn't verbally get it out," according to Mrs. Rosen.

Quick to Read, Slow to Talk   Alex didn't say his first word until three and a half years of age and didn't initiate conversational exchanges until almost four. His enthusiasm for reading, in contrast, progressed at a rapid pace.

"Since about age six, reading has become almost an obsession," according to Eileen. "He would rather go to a bookstore than go to a toy store. I could leave him at Borders or Barnes & Noble and come back three hours later and he'd still be in there reading. He has wide-ranging interests that include history, science, geography, geology, and weather patterns. And as a result of his reading so much, he has an incredible number of facts in his brain."

At the time of our meeting, Alex was eight years old and in the second grade of his local elementary school. He could read newspapers (most newspapers are written at about a sixth-grade level) and, in contrast to many hyperlexics, he could comprehend most of the material he read. His mother told us about an experience that had occurred a few days earlier:

"When Alex was introduced to his new principal, he read the diploma on her wall aloud to her. The diploma was written in an archaic cursive, to boot. So now we know he can read both cursive and print despite the fact he hasn't yet learned to write in cursive."

As a general rule, most hyperlexics can be placed somewhere along a spectrum consisting of autistics at one end and at the other end children who have problems with spoken language and communication. Where does Alex fit on this spectrum? At first glance Alex's "obsession" (his mother's term) with reading, along with highly unusual interests for an eight-year-old (the British monarchy and Henry VIII) might suggest that Alex is autistic. But one has only to spend a few moments with this child to discard any suspicions of autism.

"His teachers adore him because he relates very well to adults," his mother responded with pride and gratitude. "Everybody who knows him agrees that he's a very caring, warm kid."

Since Mrs. Rosen had taken Alex to Guinevere Eden for an evaluation, we asked her about him and hyperlexics in general.

"Hyperlexic readers are so unusual because, unlike the rest of us, they don't have to be taught how to read. Instead their difficulty involves speaking and understanding the speech of others. When I first saw Alex at age six he was reading at the level of a twelve-year-old. Now at age eight his reading ability is somewhere between a fourteen-year-old and an adult."

Testing Pattern In order to test Alex's precocious reading ability Eden placed him in an fMRI scanner. She found that his activation pattern looked similar to highly skilled adult readers. "That is very unusual," says Eden. "His brain clearly is wired very strongly to serve reading. That's his particular skill and he's very good at it."

By studying Alex and other hyperlexics, Eden hopes to understand how and why hyperlexics can read so much earlier and so much more skillfully than they can talk. "Many of the hyperlexic children are like Alex and display delayed spoken language. We want to get a better sense of the relationship between reading and spoken language and how disturbances in this relationship lead to language delay."

Eden also hopes that an understanding of brain functioning in hyperlexics will shed light on the processes in normal readers. Does the hyperlexic child use the same pathway as an older child who reads at the same level? Researchers suspect that different mechanisms are involved. If this turns out to be true, such findings might provide information about other areas of the brain that dyslexics can call upon to compensate for their reading difficulty.

Researchers' suspicions are strengthened because fMRI images from adults who participated in a remedial program showed changes not just in the areas of the brain known to be important in language but also in the motor and visual areas, along with other areas known to be important in skill acquisition.

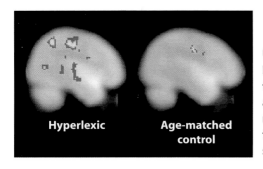

Hyperlexic   **Age-matched control**

Unlike other children, hyperlexics do not have to be taught to read. As seen at left, the brain of a hyperlexic reader is considerably more active than that of an age-matched control. In some cases, the activation pattern looks similar to that of a highly skilled adult reader.

"And that shouldn't be surprising, really," comments Eden. "Reading is a form of skill acquisition, and in our fMRI studies we see activity in areas known to play a role in learning complex skills."

In summary, it seems fair to say that neuroscientists have learned a lot in the last decade about what's happening in our brain when we're reading. The largest contributor to this enhanced understanding has come from advances in neuro-imaging. For one thing, language is a time-sensitive process: We learn languages more easily when we're younger. This isn't absolute, of course. It's never too late to learn a new language. But the learning will be harder, and it's unlikely that we'll ever completely master the grammar no matter how many new words we learn. Thus time sensitivity and plasticity operate as two separate but mutually interacting processes within the brain.

While the brain remains plastic and capable of change up until our final years, time-sensitive periods exist for language. And as the brain further matures, other time-sensitive processes come into play, placing additional limits on plasticity.

The seemingly unlimited plasticity of the child's brain is lost as we humans pass from childhood into adolescence. But there is one area that remains highly plastic—the region responsible for our powers of reason and self-control. The transition from childhood to adolescence involves opportunities as well as threats, but it is the next great step that the developing brain must take.

FROM CHILDHOOD TO ADOLESCENCE

An adolescent's thinking differs from that of a child's in the adolescent's capacity to form the question: What if? In contrast to the child who is locked into the here and now, the adolescent can think in terms of imagined possibilities, what might be true rather than simply what is true at the moment. While children usually remain focused on the things and people around them, adolescents brood about complex problems, analyze subtle moral dilemmas, and envision ideal societies.

Adolescence is marked by a melange of idealism, romanticism, theatricality, self-indulgence, judgmentalism, and disillusionment. And because adolescents can imagine a rich variety of possibilities, they're impatient to take steps toward trans-forming their images of the future from concept into reality. As a result of this impatience, adolescents traditionally impress their elders as needlessly rebellious. The adolescent may argue with parents and other authorities, overreact to criti-cism, respond with indecisiveness when required to make decisions, and fail to rec-ognize the key difference between speaking of an ideal and taking practical steps toward its achievement. The result is intergenerational conflict as depicted in such classics as "Romeo and Juliet" or, more recently, "Rebel Without a Cause."

Traditionally, the changes that accompany adolescence are considered from psychological, social, and even political points of view. But many aspects of adolescence—indeed some of the most puzzling and frustrating—result from changes that take place as the brain of the child makes the transition to the brain of the adolescent.

A World of Their Own

THE ADOLESCENT BRAIN

The *New Shorter Oxford English Dictionary* defines adolescence as "the process or condition of growing from childhood to manhood or womanhood, the period of growing up." It is the time when we learn to understand the world around us, other people, social rules, and abstract ideas, when we develop abiding beliefs about how the world functions. "Adolescence is the time when you grow into your own skin," as one adolescent described the process. But the most important aspect of that growth concerns not the skin, but the brain.

The adolescent brain occupies the middle ground. It's neither the mature brain of the adult nor the inchoate, unsettled brain of the child. Flexible and adaptable, the adolescent brain is totally appropriate for a turbulent and often painful period when everything is in flux: personality, identity, socialization, emotional control, and logical thinking.

Since the human brain reaches its full size by the age of five, some scientists used to assume that the organ was fully wired by the end of childhood, that all subsequent development involved only some fine-tuning. But those scientists were wrong.

Within the adolescent brain wide-ranging changes are occurring in anatomy, neurochemistry, and hormones. Neurons are establishing connections that will last a lifetime. Genes are changing their expression and thereby dictating the protein structure of cells in the brain and throughout the body. In short, it's a time of tremendous fluidity that perfectly mirrors the turmoil of adolescence.

A Work in Progress "We have only recently learned that adolescence is a work in progress, a distinct stage in brain development," says Jay Giedd, a neuroscientist at the National Institute of Mental Health (NIMH) in Bethesda, Maryland. Over the last several years, Giedd and his colleagues have performed biennial magnetic resonance imaging (MRI) on a group of normal children and adolescents in an effort to correlate brain changes with behavioral changes. The changes were also correlated with human post mortem findings dating back more than 20 years.

The initial MRI findings came as a surprise to Giedd and his coworkers. Even though, on average, brain size doesn't change very much, some brain areas undergo dynamic alterations. For instance, while the white matter remains about the same within the brain's various lobes, the gray matter shows an initial increase in preadolescence, followed by a decrease with the arrival of adolescence. This suggests that the brain undergoes a second wave of nerve cell production and elimination during adolescence, similar to the process that occurred during the first 18 months of life.

Until age 11 in girls and 12 in boys, brain cells grow new connections like a tree growing extra roots, branches, and twigs. Then comes the process of pruning and the overall elimination of many cells and a thinning of the brain's gray matter. (The principle underlying this selective elimination is sometimes referred to as "neural Darwinism.") In Giedd's words, "This very dynamic time of adolescence is marked by the overproduction of neurons and their selective elimination. In the first phase of the process, just as puberty begins, the brain grows an exuberance of new connections. Then through experience and learning—'use it or lose it,' as we refer to the second stage of the process—some of these connections are pruned away. And rather than a cause for regret, neuronal cell pruning within the adolescent brain is an optimistic finding since it actually makes the brain more efficient."

To get a feeling for this enhanced efficiency, picture the midtown streets of a major city clogged with traffic. Everything is congestion, noise, conflict, and

paralyzing inefficiency. In response, traffic engineers reduce the number of cars funneling into the most congested areas of the city by rerouting the flow; eliminating certain streets; and constructing overpasses, beltways, and superhighways. When the adolescent brain is pruned it becomes like that city: with fewer neurons and fewer neuronal connections but an overall increase in efficiency.

"The adolescent brain is far more flexible and adaptable than we ever realized," Giedd told us during an interview in his laboratory at the National Institutes of Health (NIH). "There's an enormous potential for change through the teen years. And this is great because those years are a time when choices have to be made and skills acquired as adolescents learn how to adapt to their environment. During the process, adolescents must learn to gain control over their sexual and aggressive impulses, adapt their behavior to the reasonable expectations of parents and teachers, accept authority, and generally get along with others."

But not all parts of the brain mature at the same time. Those areas important in emotions—the limbic areas—mature earlier than those involved in judgment, organization and reasoning. Indeed, this discrepancy between expressing feeling and thoughtful evaluation accounts for many of the teen behaviors that so dismay parents and teachers.

"The impulsiveness, the disregard for consequences, and the rapid and unexpected emotional storms of adolescence may be in part related to the immaturity of the frontal lobes of the brain," according to Jay Giedd.

The frontal lobes, especially the prefrontal lobes closest to the forehead, are often referred to as the CEO, or executive, of the brain. Larger in size in humans

The Slow Maturation of the Prefrontal Cortex

In adolescence, teens are expected to take on at least some of the prefrontal lobe functions once handled by parents and teachers. But the prefrontal lobes are still immature—and must also keep up with the hormonal tumult occurring within the body and the brain. The adolescent experiences frustration and even anger. Teens interpret parental injunctions as "nagging." Parents, in turn, complain of the "immaturity," "impulsivity," and "lack of judgment" of their adolescent children.

How early in life does the prefrontal cortex begin to exert an influence on behavior, emotions, and temperament? Attention is the first observable measure of prefrontal functioning. Infants as young as six weeks of age are capable of attending to and anticipating events taking place around them. For example, when shown pictures that appear and disappear at regular intervals, the infant can anticipate the location of the next picture by eye shifts to the appropriate location prior to the appearance of the picture. By 18 months the infant can perform complicated maneuvers requiring sustained focused attention.

Between eight and twelve months of age an infant will move an intervening object in order to grab a toy—a neat example of what psychologists refer to as "means-ends" behavior. By one year of age an infant also begins to represent the world by means of language and symbols. This provides the earliest time line linking past and present, present and future. From these links will emerge the earliest memories since, essentially, we best remember those things that we can symbolize and put into words.

Self-control (impulse control) is the third skill mediated by the prefrontal cortex. Examples include compliance with requests, toleration of delay, and conforming one's behavior to the demands of the circumstances. The most consistent increase in self-control occurs from 18 to 30 months, especially between 24 and 30 months, when temper tantrums decrease. Still, parents and teachers must act as stand-in prefrontal lobes for young children because the child's prefrontal lobes aren't yet mature enough to foresee adverse consequences.

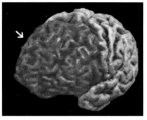
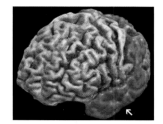
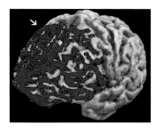

| 3 to 6 years | 7 to 15 years | 16 to 20 years |

By monitoring brain development, researchers have discovered that the brain undergoes waves of growth (red) and pruning (blue). Between the ages of 3 and 6, children experience tremendous growth in their frontal circuits, accounting for increased attention, vigilance, and alertness. Between 7 and 15, the wave of growth occurs in the temporal and parietal lobes; this period is the most efficient time for children to learn other languages. Between 16 and 20, the frontal lobe undergoes pruning, as life skills are honed. The adolescent gains greater self-control, becomes better at planning, and learns how to regulate his or her behavior.

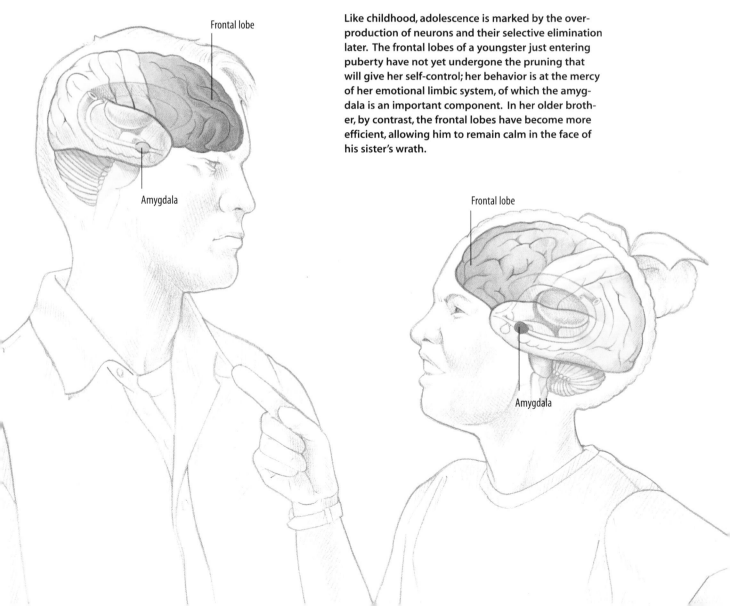

Frontal lobe

Amygdala

Like childhood, adolescence is marked by the over-production of neurons and their selective elimination later. The frontal lobes of a youngster just entering puberty have not yet undergone the pruning that will give her self-control; her behavior is at the mercy of her emotional limbic system, of which the amygdala is an important component. In her older brother, by contrast, the frontal lobes have become more efficient, allowing him to remain calm in the face of his sister's wrath.

Frontal lobe

Amygdala

The activities an adolescent chooses—be it sports, music, studying, or watching television—will shape the brain cells and connections that survive into adulthood.

than in other primates, the prefrontal lobes are the last of the cerebral areas to reach maturity. While the motor and sensory centers are almost fully developed in early childhood, the prefrontal lobes aren't fully mature until the 20's or even later. In short, the immaturity of the adolescent's behavior is perfectly mirrored by the immaturity of the adolescent's brain. Areas likely to be deficient include planning, organization, and strategizing.

In early childhood parents act as accessory prefrontal lobes because the child's prefrontal lobes aren't sufficiently matured to foresee possible adverse consequences. Parents' injunctions ("Don't cross even a one-way street without looking both ways.") provide the child with his earliest examples of cause-and-effect relationships. Teachers also provide prefrontal functions as parents by proxy.

With the arrival of adolescence, the teen must take on at least some of the prefrontal lobe functions formerly supplemented by parents and teachers. But since the prefrontal lobes are typically too immature to do a good job, the adolescent experiences frustration and even anger. Injunctions are now interpreted as "nagging."

"In adolescence the frontal lobe is asked not only to keep up with the biological tumult going on within the brain, but at the same time adapt to a more complex environment as an independent adult. It's a time when the capacities of the prefrontal lobes are severely tested," according to Daniel Weinberger, a psychiatrist at the NIMH and an international leader in schizophrenia research.

The good news is that the adolescent can play an active role in determining the structure and functioning of his or her own brain. If a teenager concentrates on math, music, or sports, the brain will incorporate these activities in the form of neuronal circuits. But if the teen spends the day just "hanging out" or passively watching television, the brain will fashion corresponding circuits for these activities as well. Thus, the adolescent brain exhibits a tremendous plasticity. Indeed the adolescent's choices determine the quality of his brain.

"This is a very empowering concept," says Giedd. "In essence the adolescents, on the basis of the activities they engage in, choose the brain cells and connections that will survive as they move into adulthood."

But what happens if the prefrontal lobes aren't functioning well enough to hold in check the emotional impulses that accompany adolescence? For one thing, such adolescents will experience difficulty in self-control. The most common difficulty is attention deficit hyperactivity disorder (ADHD).

ADHD: A Worldwide Problem   ADHD is prevalent throughout the world, with estimated incidence among children of four to five percent. Although usually starting before age seven, ADHD persists into adolescence in three-quarters of cases and into adulthood in approximately half the cases. Furthermore, the symptoms of ADHD often change during the transitions from childhood to adolescence to adulthood. Children with ADHD typically show hyperactivity, impulsivity, low frustration tolerance, and several other symptoms (see page 78).

To psychiatrist Timothy Wilens, of the Clinical Research Program in Pediatric Psychopharmacology at Massachusetts General Hospital in Boston, the effects of ADHD during adolescence can be compared to congestion on a major highway: "When a main thoroughfare is congested you have to use one of the collateral roads. Not only does this take longer, but it's also a lot less efficient. A similar process occurs in the brains of people with ADHD. Those parts of the brain responsible for attention—principally the lateral prefrontal cortices—are not operating optimally, and therefore a lot of collateral, less-efficient brain pathways have to be called into play."

The Challenge of ADHD

In today's world, where the rapid assimilation of information is essential, the inability to focus leaves adolescents with ADHD at a grave disadvantage. When their attention problems cause them to fall behind at school, they may become demoralized and depressed, which only exacerbates the problem.

If ADHD is diagnosed, long-term treatment with counseling and medication may be required, including treatment for accompanying anxiety and depression, if any. Helping children handle such strong, painful feelings will help them cope with and overcome the effects of ADHD, especially as they make the transition to adolescence.

Symptoms of Attention Deficit Hyperactivity Disorder (ADHD)

Children with ADHD may exhibit the following often-occurring behaviors:

- **Fidgets with hands or feet or squirms in seat**
- **Leaves seat in classroom or in other situations in which remaining seated is expected**
- **Runs about or expresses a subjective feeling of restlessness**
- **Has difficulty playing or engaging in leisure activities quietly**
- **Is "on the go" or often acts as if driven by a motor**
- **Talks excessively**

In addition, children with ADHD may exhibit symptoms of impulsivity, such as:

- **Blurts out answers before questions have been completed**
- **Has difficulty awaiting turn**
- **Interrupts or intrudes on others (e.g., butts into conversations or games)**

To earn the diagnosis of the "inattentive subtype" of ADHD, the adolescent has only to meet any six of the nine following symptoms:

- **Fails to give close attention to details or makes careless mistakes in schoolwork, work, or other activities**
- **Has difficulty sustaining attention in tasks or play activities**
- **Does not seem to listen when spoken to directly**
- **Does not follow through on instructions and fails to finish schoolwork, chores, or duties in the workplace**
- **Has difficulty organizing tasks and activities**
- **Avoids, dislikes, or is reluctant to engage in tasks that require sustained mental effort (such as schoolwork or homework)**
- **Loses things necessary for tasks or activities**
- **Is easily distracted by extraneous stimuli**
- **Is forgetful in daily activities.**

Based on the *Diagnostic and Statistical Manual of Mental Disorders* (DSMIV)

Wilens thinks of the frontal lobes as providing "a dampening effect, a kind of shock absorber" that enables the adolescent to inhibit rapid inappropriate responses. "When we've scanned the brains of adolescents with ADHD we've seen reduced functioning in those brain areas that are most involved in attention, concentration, and focus: the prefrontal lobes, especially the lateral prefrontal areas, and some of the nuclei located deep beneath the cortex."

As a result of these brain problems, ADHD sufferers experience difficulties with organization, time management, and priority setting. They can't decide when to start, when to stop, how to avoid distractions, and how to focus on one task at a time. Indeed, focus problems are at the core of this disorder.

"Adolescents with ADHD lack the ability to focus," says Wilens, "to concentrate on one thing long enough to extract information in an efficient way. Such failures are a big handicap because today's adolescents often need to rapidly scan a lot of information, assimilate what's important, and focus on those areas that are going to help them achieve a particular goal."

As a result of these attention problems, the adolescent afflicted with ADHD starts falling behind in class and can't keep up with even less gifted students. In response, chronic frustration merges into demoralization, even clinical depression.

In search of relief from frustration and depression, the adolescent with ADHD may—as we will discuss in more detail later—turn to drugs of abuse as a means of combating their troubling symptoms.

Twice the Risk of Substance Abuse   "Some adolescents with ADHD use drugs to dampen internal feelings of restlessness, anxiety, and poor self-image," according to Wilens. "Others use drugs like cocaine or amphetamines in order to help them focus. But whatever the drug or whatever the reason for using it, adolescents with ADHD are at twice the risk of developing a drug or alcohol problem through adolescence and into young adulthood."

This increase in substance abuse in adolescents with ADHD is also found among their family members who are not affected by the disorder. Conversely, within families with a high rate of substance abuse, the incidence of ADHD is also increased. Does this mean that adolescents with a genetic risk for ADHD may also carry a greater genetic risk for addiction of some kind?

"We think that ADHD and substance abuse are associated and it may be that if you have ADHD by genetic risk—about 25 percent of the cases—you're also at increased risk for developing a substance abuse problem, typically starting during adolescence."

ADHD also carries an increased probability of developing conduct and mood disorders such as depression (20 to 30 percent). Conversely, high percentages of ADHD are observed in depressed (20 to 30 percent) and bipolar youths (50 to 90 percent).

To Medicate—or Not?   Treatment for ADHD remains controversial. While recent research suggests a deficiency in dopamine functioning, many parents and some educators remain unconvinced that ADHD should be treated with medications. Yet a very powerful argument can be made for treating ADHD with medications. Medication-treated adolescents with ADHD have a 66 percent reduction in risk for the subsequent use of alcohol, cocaine, stimulants, and other drugs. This difference between the two groups may be due to the mechanisms of action of the most commonly used and most successful medications used for treatment.

Psychostimulants like Ritalin and Dexedrine are currently considered the "first-line" agents for ADHD. They work by increasing the concentration of dopamine and other neurotransmitters within the synapse. This is accomplished

Though the biological cause of ADHD remains unclear, most research points to insufficient dopamine action in the frontal cortex and striatum areas of the brain. ADHD medications such as Ritalin (methylphenidate, or MP) target the dopamine system. The scans at right highlight the availability of dopamine receptors (red) after administration of a placebo or Ritalin. In the striatum (near right), where Ritalin competes with dopamine for receptors, fewer receptors are available with Ritalin (bottom) than with the placebo (top). By occupying dopamine receptors, Ritalin forces more dopamine to remain available to the brain. The effect is not seen in the cerebellum (far right, top and bottom), where binding is less specific.

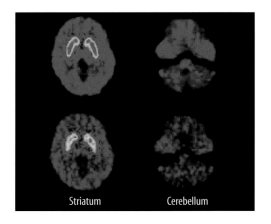

Striatum   Cerebellum

either by inhibiting the reuptake of the neurotransmitters or releasing additional amounts of the neurotransmitters into the synapse.

Recent positron-emission tomography (PET) studies show that Ritalin increases dopamine levels in the brain. It's speculated that the increased level of dopamine produced by Ritalin may amplify dopamine activity between neurons and thereby lead to improved concentration.

But the research on stimulant use for ADHD is also providing critics with a new objection to the use of medications like Ritalin and amphetamine. Since many of the drugs of abuse, such as cocaine and methamphetamine, also increase dopamine within the synapse, the critics charge that the use of stimulant medications is merely a legalized, socially acceptable means of achieving the same result as when adolescents buy the same drugs off the street.

"Despite superficial similarities, there are a lot of important differences," according to Wilens. "Both the doses and the routes of administration are different. Smoking or shooting amphetamine involves greater amounts of drug and more widespread effects on the brain. If the amphetamine is prescribed and taken therapeutically, the uptake into the brain is different. And you really can't compare cocaine to any of the drugs we currently use for ADHD. They bind entirely differently to their receptors. Comparing them is like looking down from a satellite at two cars and saying they're exactly the same when closer inspection reveals that one is a Ford Escort and the other is a Mercedes Benz. Same cars? No way!"

Experience with ADHD has accustomed researchers to think of the prefrontal lobes as a braking mechanism that acts by encouraging the adolescent to think such thoughts as, "Wait a minute, I don't think I'll like the consequences that'll come if I take the family car without asking permission." For a subgroup of teens, the braking mechanism is malfunctioning; their frontal lobes remain underdeveloped in comparison with other teens.

"The brakes may not be as good in a vulnerable subgroup of adolescents," according to Anna Rose Childress, a research psychiatrist at the Treatment Research Clinic at the University of Pennsylvania, in Philadelphia. "They are particularly vulnerable if they encounter the compelling rewarding and pleasurable experience associated with drugs."

The Dark Side of Plasticity   Drug use is the dark side of the plasticity and dynamism of the adolescent brain. In too many instances the adolescent's first encounter with drugs comes at a time when the prefrontal fibers are not yet mature—the braking system only precariously established. After hearing about drugs, reading about drugs, and observing their effects in others, the adolescent may give into peer pressure and decide to try them.

Many people consider drug use an expression of a "weak character" or a failure of will. For such people, the adolescent who becomes addicted to drugs has nobody to blame but himself for his addiction. But several facts argue against such a harsh and strictly moralistic interpretation.

For one thing, many of the adolescents who become addicted to drugs didn't exhibit conspicuous character flaws prior to their first experience with the drugs. They didn't steal or lie or display other signs of weak character until they came under the influence of the drug.

In the second place, drug use is more common among adolescents with neuropsychiatric disorders such as ADHD, depression, and anxiety. All of these conditions are now recognized as brain disorders, not personality or character flaws.

Finally, addiction or dependence on drugs often runs in families. One member of the family may have an alcohol problem; another member grapples with addiction to heroin; a third member can't seem to stop smoking despite vast expenditures of time, money, and effort.

In essence, drug addiction is a brain disease. We know this because contemporary neuroscientific research reveals that addiction is based on the power of drugs of abuse to mimic the action of one or more of the brain's neurotransmitters. As a result of this successful mimicry, the brain is fooled into responding as it would to the natural neurotransmitter. Most important, the drug of abuse acts in areas of the brain collectively referred to as the "pleasure centers."

The Reward Pathway   Neuroscientists have known for years that certain areas of the brain spring into action whenever we experience pleasure. In 1954, James Olds and Peter Milner showed that it's possible to enhance learning in rats by electrically stimulating their brains. But in order to be effective, the stimulation must be delivered along a specific pathway named the "mesolimbic reward system."

This pathway extends from the brain stem upward to specific sites deep within the brain. Included within this circuit are components with exotic names like the nucleus accumbens, the septum, and the ventral tegmental area, among others (see page 84). But despite vast differences in connectivity and function, dopamine is the neurotransmitter common to all of them. Increasing the level of dopamine increases neuronal firing in these areas; decreasing the dopamine level decreases the firing rate.

As an example of the relevance of this finding to daily life consider yourself as you bite into and savor a piece of strawberry cheesecake (assuming you like cheesecake). At that instant, neural messages zip from regions in your brain stem to the septum, nucleus accumbens, parts of the frontal cortex, and many other targets within the dopamine pathway. If electrodes could be inserted safely into those areas of your brain and the results displayed on a monitor, you could watch the equivalent of an incredibly scintillating game of pinball as the pleasure centers light up and play off one another in response to the cheesecake.

If you don't like cheesecake, think about whatever else "turns you on"—a term, incidentally, that provides an apt description of what's happening in your brain when you're experiencing pleasure. Whether it's movies, skiing, travel, parties, or whatever, the brain's response is the same: activation, "turning on," of those pleasure centers.

Unfortunately, these same areas are also turned on by drugs of abuse. We learned this, too, from early experiments with rats. When laboratory rats eat or become sexually aroused, dopamine levels increase in the areas listed above, most notably in the nucleus accumbens. But the same thing happens when rats inject themselves with addictive drugs by pressing a lever or turning a wheel. They will repeat these behaviors hundreds of times in order to experience the pleasure set off by cocaine.

A similar frenzy was described for us by Liz, a patient at the Caron Foundation in Wernersville, Pennsylvania, that specializes in the treatment of cocaine and other substances of abuse.

"When I started getting high, I loved it more than anything in life. It started to be the most important thing in my life. I didn't feel normal if I wasn't high."

Pleasure often seems to come at a price, doesn't it? Think back to the cheesecake. Along with a good deal of pleasure, that piece of cheesecake also

The Reward Pathway Deep within the brain are several structures that evolved to respond to pleasurable stimuli, such as food and sex, with a cascade of neurotransmitters, chief among them being dopamine. The cascade begins when neurons in the hypothalamus release serotonin. This triggers release of other neurotransmitters that in turn allow cells in the ventral tegmental area (VTA) to release dopamine. The dopamine travels to the amygdala, the nucleus accumbens, and certain parts of the hippocampus.

Along with the prefrontal cortex, these structures are involved in memory formation, explaining in part why visual and other cues associated with drug use can trigger intense cravings in people addicted to such substances as alcohol, nicotine, and cocaine. By hijacking the reward pathway, these drugs make addiction very difficult to overcome.

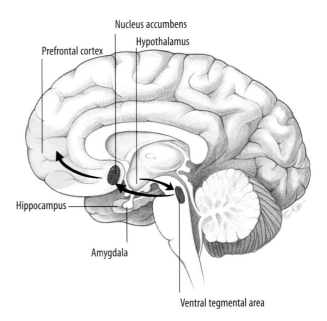

can provide you with enough calories to force you into shopping for another bathing suit. Indeed, if continued long enough and intensely enough, any of the things that give pleasure eventually exact a cost. The same thing holds for drugs.

As Liz tells it: "It got to the point when I couldn't even form sentences anymore. I had bags under my eyes. And my room—like everything around me—was a wreck and just falling apart. Without my drug I would be impossible to be around. I'd be always angry and snap for absolutely no reason. I would just hate myself for acting like that."

Who's At Risk? But why can some adolescents try drugs and walk away from them, while others, like Liz, become hooked? According to Steve Hyman, the director of the NIMH, "An adolescent at risk for addiction starts off with something like attention deficit disorder, depression, or perhaps just a difficult temperament. As a result, the healthy, well-socialized peer group doesn't like them; their teachers don't like them; even their parents can barely tolerate them. In response, the adolescent becomes disgruntled with school and with the future in general. He starts hanging out with an antisocial group who introduce him to drugs."

And, according to Hyman, the process of social isolation only gets worse with drug use. An adolescent addicted at 14 misses out on the normal, healthy development of the prefrontal cortex that ordinarily provides the executive function, the moral dimension, and the sense of personal responsibility. The addicted adolescent also misses out on the normal day-to-day communication that ordinarily goes on with family and friends.

Disturbed social communication in addiction is mirrored in the brain at the chemical level by neurotransmitter disturbances. When a person takes an addictive drug, a massive surge of dopamine takes place within the pleasure circuits. This surge is far in excess of what occurs when the person experiences a normally positive event like winning in a sports competition or gaining a promotion at work. The surge is so great that it even eclipses the pleasure of orgasm. The message transmitted to the rest of the brain from the pleasure centers is: "This is the most rewarding, exciting, important moment that you could possibly have as a human being." Indeed the experience is so overwhelmingly significant that it builds within the structure of the brain the overwhelming desire to repeat it again.

Trojan Horses   As Hyman explained: "All addictive drugs are Trojan horses that mimic natural neurotransmitters or affect those neurotransmitters. The drugs do that by tapping into the brain-reward pathways and fooling them. As a result, the reward pathway signals, 'That felt good. Let's do it again. Let's remember exactly how we did it.' So when you put a drug of abuse in your system, that drug is instructing that system, 'Here is something important, something to pursue, something you're going to remember.' As a result you're going to repeat that experience again and again. That compulsion to repeat the drug experience is the barb in the hook that snares the addict."

While addiction is difficult to overcome at any time in a person's life, it's especially difficult during adolescence. Because of the incompletely developed prefrontal lobes, that all-important braking mechanism isn't in place. Thus, consequences can't be foreseen, and momentary pleasure trumps any thought of future consequences.

During adolescence, when the shape and functioning of the brain remain in flux—some circuits strengthening, with others disappearing—exposure to drugs of abuse can create a situation where pleasure responses essentially consist of drug-associated electrochemical firing of the reward pathways of the brain. And as a result, adolescents who are abusing drugs lose out on the richness of human experience.

Ecstasy: The Popular Brain Killer

MDMA, known as Ecstasy, first hit the streets in the 1980s and has since grown exponentially in popularity. (The National Institute on Drug Abuse estimates that 11 percent of twelfth-graders have tried it.) Ecstasy stimulates the release of serotonin in the brain, producing a euphoric high and weak, LSD-like effects that can last several hours.

But Ecstasy is also neurotoxic, causing lesions at the terminals of neurons that produce serotonin, destroying nerve cells and affecting areas of the brain responsible for learning and memory. Studies have shown that the damage caused by Ecstasy persists up to seven years after its use, including permanent damage to the neurons that produce

Nor is this their only loss, for adolescents hooked on drugs like cocaine are cosignatories in a devil's bargain. They constantly yearn to reexperience the initial feeling they had with cocaine, but they discover that over time, increasing amounts of cocaine are needed to get anywhere near that initial feeling. According to addiction treatment specialists, the drug-addicted adolescent never gets that feeling back. But he or she keeps trying, nevertheless. Additional drugs are taken in a vain effort to recapture an experience that seemed so wonderful at the time.

A Hijacked Brain Not surprisingly, the treatment of adolescent addiction isn't easy. Why give up the possibility—however elusive—of recapturing that initial "rush" that accompanied the first few drug-taking experiences? How can you "just say no" when the drug of abuse has literally hijacked the brain's structure and functioning? And, as a final kicker, the adolescent brain exacts a harsh and painful penalty for the drug user who decides to back out of that devil's bargain.

Eventually, the adolescent brain becomes wired for both the pleasure initially associated with drugs and the anticipation of drugs. As a result, the adolescent becomes attuned to people and events usually associated with the drug experience. Merely returning to the neighborhood where drugs are available sets off an intense craving. Even if the user doesn't take the drug, he or she experiences many of the actual effects of the drug. The heart rate picks up, tingling can be felt in the head or other body parts, a buzzing sensation is experienced along

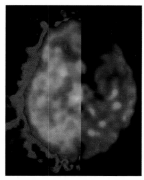

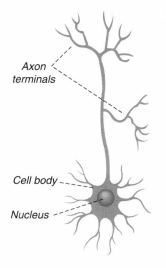

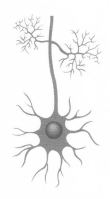

A composite fMRI scan (left) of a drug-free brain (left side) and a brain under the influence of MDMA (right side) reveals how Ecstasy, the popular "rave" drug, dramatically alters cerebral activity levels. The illustration at right shows how, 12 to 18 months after being damaged by MDMA, nerve fibers in the brain have regrown excessively in some areas and not at all in others.

Axon terminals

Cell body

Nucleus

Normal   Long-term effect

serotonin. Habit-forming and subject to compulsive abuse, Ecstasy increases blood pressure to dangerous levels. Long-term use can cause sharp increases in body temperature, leading to kidney and liver failure.

with light-headedness and, in many instances, even a bit of the drug-associated euphoria.

So when an adolescent who has become hooked on cocaine in the recent past walks down the street and sees a little plastic vial in the gutter, this vial—which to anyone else is just a vial—sets off a whole reverberating circuit of activity in the brain. Anna Rose Childress, the University of Pennsylvania psychiatrist who specializes in the treatment of drug addiction, says: "The brain signals the message, 'Cocaine is somewhere near; I could get this cocaine with a little bit of effort and luck, or maybe a little bit more money than I have at the moment.'"

In response to this intense craving, the adolescent may do things he would have never dreamed possible. Stealing and lying may become second nature as a result of the profound personality transformation brought on by the drug.

A Stranger in the Mirror "This personality transformation involves crossing moral and behavioral lines," Childress says. "Since, in order to get drugs he may engage in illegal activities that are against his moral code, the adolescent becomes a stranger to himself, someone he no longer recognizes. And accompanying this loss of identity is a sense of shame and guilt. The adolescent's world is literally turned upside down; friends, family, jobs, cars—nothing even compares to the importance of the drug."

THE ADOLESCENT BRAIN

To understand the brain-basis of addiction and "peek inside the brain of the addict during desire," Childress has looked at PET scan recordings of addicts. After placing the subject inside the scanner, she projects on a small screen a videotape of either non-drug-associated scenes (Childress uses pictures of hummingbirds) or scenes of people simulating the buying, using, and exchanging of cocaine. She's found that the cocaine-related tapes not only induced a craving for drugs, but she could detect "a clear signature for desire" on the PET scan. The anterior part of the cingulated gyrus, the amygdala, and the nucleus accumbens—three key way stations in the mesolimbic pleasure-reward system—lit up. This finding was consistent with previous animal studies that showed a gush of chemical messengers—dopamine principally—occurring in these sites after cocaine administration.

But no activation occurred when the addicts looked at the hummingbird videos. Nor did Childress detect any activation when a control group of normal volunteers watched either the hummingbird or the cocaine-related videos.

Another study of cocaine users showed a marked overlap between activation patterns during cocaine-induced euphoria and the patterns associated with looking at or hearing a loved one. This suggests that the pathways responsible for the pleasure response to drugs like cocaine evolved from the neural systems for social attachment. If true, this would help explain why drug habits are so difficult to break.

Eighty Percent Relapse   So powerful is the craving for drugs that 80 percent of those who have completed drug treatment programs relapse within six months. In the vast majority of those drug treatment failures, a powerful craving precedes the return to drug use. This sequence occurs even among seemingly highly motivated and committed adolescents. In essence, craving wins out over resolve. To Dr. Childress, the battle is between what she calls the "inexorable pull" back to drugs and the "stop systems" supplied by the frontal lobes.

"In adolescence you have a brain that is exquisitely sensitive to pleasures and rewards but the stop systems, the braking systems, aren't yet in place and may be vulnerable to assault from these highly rewarding chemicals."

Unfortunately, there is currently no drug that on its own can be reliably depended on to control drug abuse simply by altering dopamine or its effect in the pleasure circuit. The best approach at the moment involves bolstering the action of the frontal lobes.

According to Childress: "In this struggle between the Big Go and the Big Stop regions we look for every way of enhancing the latter. But it is so difficult because the Go system is so powerful—like a snowball rolling down a hill. So we focus on therapies that detect that first hint of a Go response, the first hint of craving. We train the adolescent to bring into play his coping strategy based on asking himself, What will be the consequences? We get him to put the brakes on early."

As Childress points out, any successful treatment must prepare the addicted adolescent for the craving that will emerge on return to the previous environment. On completion of a course of treatment, most adolescents feel confident that drug use is behind them. But their most difficult challenge lies ahead.

Images of Desire One of the biggest obstacles to overcoming addiction is the powerful craving an addict experiences when exposed to drug-related cues. A cocaine addict watching a video showing images of cocaine use experiences heightened activity in brain regions involved in aspects of memory and learning (below, far left), including the amygdala, temporal pole, and the anterior cingulate in the prefrontal cortex. Dr. Anna Rose Childress at the University of Pennsylvania is studying the use of a common antispastic medication called baclofen to dampen drug cravings. As the scans at near left indicate, a recovering

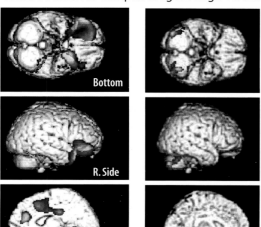

Bottom

R. Side

Middle

cocaine addict (below) who is taking baclofen experiences very little activity in those regions when exposed to a video showing drug cues.

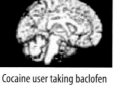

Untreated cocaine user   Cocaine user taking baclofen

While visiting the Caron Foundation we observed firsthand the difficulties involved in formulating a successful approach to getting adolescents off drugs. In our interviews we came face-to-face with the destructive effects inflicted on adolescents by addicting drugs. This is what Rick told us on day eight of his 28-day treatment program:

"I enjoyed being high. When I wasn't high I hated my life, I despised it. Besides, when I was off drugs I was irritable—all I could think about was how I was going to get more drugs."

"Do you want to beat your addiction, Rick?" we asked him.

"That's a hard question to answer. I loved being high and I always will. But I don't want it to run my life anymore. I don't want to be stealing from people in order to get high. I don't want drugs to be affecting my brain and my life. I would rather have it be like a hobby. But if I have to give it up, if it has to be all or nothing, then I guess I do want to beat my addiction."

Rick's tentativeness and irresolution suggests a bad prognosis for recovery unless he starts thinking of drugs as a problem rather than a solution to his depression. And even then, the process of recovery may be complicated by genetic factors: Rick's father was an alcoholic; an uncle, a heroin addict, died of AIDS contracted from the use of contaminated needles. Rick worries about the genetic connection:

"The loss of my uncle hit me hard because I've been hearing all my life how much I look like him and talk like him. And since my father was an alcoholic, I swore I was never going to be like him—I wasn't going to be a drunk. But my addiction has caused similar pain and problems for my family. I just substituted drugs for alcohol."

Liz started smoking pot while attending a private school. In her own words "she wanted to feel normal." She compares the experience to smoking a first cigarette: "You put it in your mouth and you take a drag and your brain tells your body, 'This is not good for you.' And in response, your body coughs and tries to get it out. But then as you start doing it more and more, your body gets used to the nicotine and the body then says to the brain, 'I need this. I'm running out of nicotine. I need more nicotine.' "

But at this point in her treatment program Liz stands a better chance than

Rick does of freeing herself from the grip of marijuana addiction. That's because she desperately wants to change. In addition, Liz has incorporated knowledge learned in a brain science class conducted by her psychiatrist about the effects of addicting drugs on the brain. While she is quietly confident of her ability to conquer her addiction, Liz also fears the "triggers":

"I have a lot of fear about those triggers. If I go home and I'm able to stay away from those people, places, and things associated with drugs, then I know I'm going to be all set. Sure I'm going to miss all my using friends, but I can't be around them. I can't be around anybody or anything that's going to get me thinking about using drugs again."

Alcohol: Legal and Pernicious   While cocaine, heroin, and pot are the major drugs of abuse encountered in treatment centers like the Caron Foundation, other perfectly legal chemicals like alcohol exert pernicious effects on the adolescent brain.

At the Veterans Administration Hospital in San Diego, California, genetics researcher Marc Schuckit has been monitoring a variety of factors in a long-term study to determine who is likely to develop alcoholism. One might expect that those people most affected by alcohol—who feel tipsy after one or two drinks—would be the ones most likely to develop problems with it. But Schuckit's findings have led him to just the opposite conclusion.

Starting in the 1970s, Schuckit compared sons of alcoholics with controls—male counterparts who were similar in age and other characteristics but with no family history of alcoholism. He studied 453 men in all, all about 20 years old, who were drinkers but not alcohol dependent.

In the lab these volunteers were given the equivalent of about three to five drinks and then underwent various physiological tests. They also filled out a checklist every half hour to rate how "high" or sleepy or nauseated they felt during each of those half-hour intervals.

The differences in response were dramatic, Schuckit found. Some of the participants in the study became uncoordinated or sleepy; their speech began to slur. In addition, these same volunteers couldn't stand for more than a few seconds before they began to sway.

By contrast, 40 percent of the sons of alcoholics (and only five percent of the controls) showed remarkably little impairment in their coordination and few body changes. For example, there was some alteration in brain waves, but it was less than expected. Asked to rate their level of intoxication, these young men would say they didn't feel drunk at all. And certainly their bodies were confirming that assessment.

Family History + Low Response = High Risk In follow-up interviews a decade later, Schuckit found that, among the men with a family history of alcoholism, those who had shown a high sensitivity to alcohol years earlier had a 15 percent risk for alcoholism. This is greater than the 10 percent risk for males in the general population, but it was nothing compared with the results for the men who as adolescents had prided themselves on being able to "handle their

Alcohol and the Brain Magnetic resonance images show significant differences between the brains of nonalcoholic women and women with alcoholism. For one, as seen in scans of the back and middle of the brain (top pair and middle pair), the ventricles of the woman with alcoholism are larger and the corpus callosum (just above the ventricles), which connects the two hemispheres of the brain, is thinner. The bottom pair of scans reveals significant shrinkage of the frontal lobes. In all three scans, more cerebral spinal fluid (CSF) surrounds the brain of the alcoholic than the nonalcoholic woman, appearing as dark inroads on the brain matter. Increases in CSF and in the size of the ventricles indicate global brain shrinkage that can impair memory and other cognitive functions.

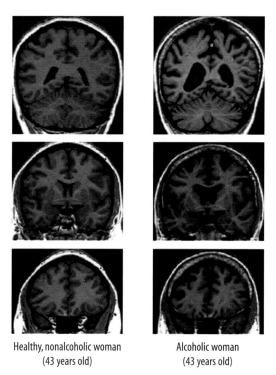

Healthy, nonalcoholic woman
(43 years old)

Alcoholic woman
(43 years old)

liquor." Those who had shown the lowest response to alcohol as youths had a 60 percent risk of developing alcoholism.

In short, some future alcoholics respond differently as an adolescent to alcohol compared to their peers who go on to develop normal patterns of alcohol use. As a result of their low alcohol response, these adolescents must take greater amounts in order to get the "high."

"This finding shouldn't have surprised me," says Schuckit. "For over 30 years I've been asking alcoholics, 'Tell me what it was like when you first started drinking.' And they typically told me that from the beginning they could drink anybody else under the table. They would say things like, 'I could drink six beers and after hours of heavy drinking be okay while my buddies who drank two or three would be pretty intoxicated.' "

In a sense, tolerance, craving, and alcohol abuse in general represent another dark side of the plasticity of the brain. The brain has adapted—in this instance maladapted—to the presence of alcohol as a necessary component to its normal functioning. It has altered its physical structure and functioning to accommodate a chemical that, in excess and over a sufficient period of time, has the potential to destroy it.

Note that Schuckit's findings are not predictive of alcoholism in specific individuals. We're talking here about vulnerabilities, not inevitabilities. A person may inherit an increased vulnerability for alcoholism just as he may inherit an increased vulnerability to developing heart disease, diabetes, or cancer. And that knowledge about potential vulnerability can be liberating rather than confining. That's because knowing about vulnerabilities provides the person at risk with the opportunity to take specific steps to minimize the chances of coming down with disease. As we will see at other places in this book, genetic predisposition is not destiny. Usually the genetic predisposition must be coupled with some environmental trigger. While a genetic "load" for alcoholism increases a person's risk for alcoholism, genetic predisposition can have no effect on a person who chooses not to drink.

Playing Russian Roulette "It's always a combination of genes and environment that make a particular adolescent vulnerable to addiction of any sort," says Steven Hyman. "Genes and environment work together like a thermostat to determine a person's responsiveness to substances of abuse. They also play a separate role in whether or not adolescents get captured by substances of abuse after they explore them. Nobody starts off saying, 'I want to become an addict or an alcoholic.' Instead, they come out the loser in a game of Russian roulette. Something about their genes, their environment, and their temperament made them vulnerable. While other adolescents often got off scot-free, they got hooked because there was something different about their brains."

Obviously, the majority of adolescents do not become addicted to drugs. Their brain develops normally; the pleasure centers respond to real-life events and people, rather than just to chemicals. Under most circumstances, the adolescent brain, thanks to its inherent plasticity, goes on to acquire the social and academic skills that can provide life's greatest pleasures: making friends, expressing and experiencing affection, living up to one's full potential. But in order for that to take place, thinking and emotions must work in harmony.

Sabrina's Story   Sometimes brain diseases that interfere with emotions and thinking can compromise the plasticity of the adolescent brain. For instance, as neurons take their place in the developing brain and join with one another in the vast intricate network of connections that make up the mature human brain, the possibility for errors arises in the form of crossed wires or defective neurons. Moreover, the resulting damage may lie dormant until adolescence. At that period, problems may suddenly arise during the fine-tuning of the prefrontal cortex that is taking place.

Sabrina Yeskel is 14 years old and lives in Raleigh, North Carolina. At age 13 she experienced the onset of schizophrenia. She describes the onset of her illness as follows: "I heard a voice for the first time while I was at camp. I didn't know at first what I was hearing. It sounded just like somebody talking to me, so I listened. At first I thought it was one of my friends talking to me. But then I realized it was in my head. Soon there were other voices. Since I didn't really understand what was going on, I tried to ignore the voices. But they kept getting louder and louder. They said I should kill myself; that my life isn't worth living. The voices put me down, calling me 'stupid,' 'ugly,' and 'fat.' The experience is like having somebody yell at you and the feeling created in your body frightens you."

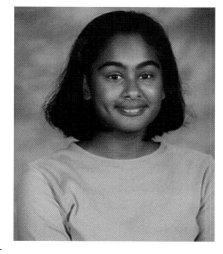

Her anxiety at going off to a sleep-away camp at age 13 may have triggered the onset of schizophrenia for Sabrina Yeskel. She began hearing voices and suffering hallucinations.

Sabrina's mother recalls the event as follows: "When I arrived at the camp to pick her up I didn't recognize her. She had been crying a lot and her face was kind of distorted. When we got in the car she started telling me that she heard her own voice in her head. By the time we drove from the camp to the motel she said the voice was different from her own. Still later it was three or four voices. Things just escalated from that point on. By the time we got home she told us the voices were telling her to hurt herself."

In addition to the voices Sabrina regularly sees a man dressed in black and wearing a black hood over his face. "Right now I see the man in the corner of my eye. He's standing there with his arms folded and with those cold eyes of hate staring at me. He's always beside me, listening to me. One time he tried to kill me and for a week after that I actually believed that I was dead."

Sabrina's mother says, "There are times when she recognizes that the man is not real. But a short time later she feels that he is real. At other times she can't decide. When Sabrina explains it to me, she compares it to a waking dream; at the moment when you wake up in the middle of the night and you're not sure what's real and what's the dream."

Schizophrenia affects one percent of the population worldwide. In 1990, the socioeconomic impact of the disease in the United States alone was $32 billion. Typically the illness appears in late adolescence in young men and in the early 20's in young women. About one person in 3,000 is treated for schizophrenia in any one year in the United States.

Patterns of Inheritance While patterns of inheritance support a strong genetic component to the illness, genes alone do not determine a person's developing schizophrenia but only increase the likelihood. For instance, a monozygotic (identical) twin of a schizophrenic has a 40-times greater chance of developing the disorder than a dizygotic (nonidentical) twin. But this leaves more than 50 percent of identical twins free of the disease. Thus, environmental factors must play a part, but what factors, and how do they come into play? To answer those questions, neuroscientists are searching for an understanding of what goes wrong in the brain that results in the illness. Take hallucinations, for instance.

Sabrina's psychiatrist, Lynn Sikich—although she readily admits she doesn't know "exactly how hallucinations are generated"—believes that they result from spontaneous stimulation within those parts of the brain that ordinarily process and interpret external sensory information. For instance, a spontaneous signal in the auditory cortex, after transfer to the auditory association cortex, will be experienced as a sound "out there" rather than as something originating within the schizophrenic's own brain. A similar process of spontaneous stimulation underlies delusions.

Spontaneous Brain Activity   Lynn Sikich: "A false perception may set off the delusion. For instance, I have a paranoid patient who arrived at the conclusion that his nurse was injecting him with something. He arrived at that delusional belief on the basis of a subtle awareness of someone coming to the door of his room. Of course, he was right: There was someone coming to his room. But that someone was the nurse who came to look in and check on him every 15 minutes to make sure he was doing well."

A psychosis such as the schizophrenia Sabrina suffers from frequently involves spontaneous brain activity that is wrongly interpreted, Sikich says. "They see or hear things that aren't there or misinterpret what's heard so that it takes on an ominous meaning. For instance, the sound of a fan going on might be heard and interpreted as voices saying 'You're a bad person' or 'Kill yourself, kill yourself.' "

From these comments by Sabrina, her parents, and her psychiatrist a profile emerges of an illness marked by delusions (Sabrina's shifting beliefs about the reality of the threatening man dressed entirely in black); hallucinations (the voices); disorganized speech and behavior (the desperate attempts to escape from her hallucinatory tormentors); social withdrawal; and disturbances in her affect, that is, feelings or emotions. Since Sabrina's schizophrenic perceptions and beliefs aren't reflective of external reality, she can also be described as psychotic: out of contact with reality.

The Role of Heredity

Although genes alone do not determine whether a person will develop schizophrenia or other disorders such as addiction, patterns of inheritance suggest that such disorders do have a strong genetic component. One of the more well-documented family histories of inherited mental afflictions is that of the Victorian poet, Alfred, Lord Tennyson, whose pedigree can be traced back to the late 17th century. As can be seen in the chart below, over time more and more people in the family were affected, and with increasing severity.

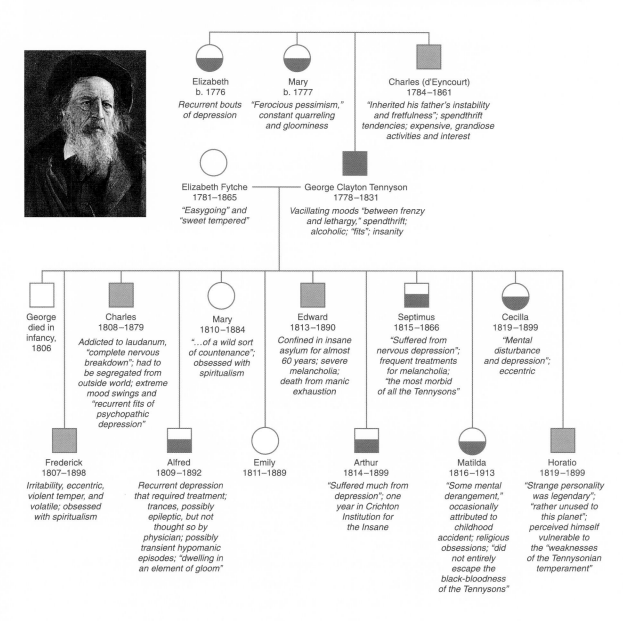

Elizabeth
b. 1776

Recurrent bouts of depression

Mary
b. 1777

"Ferocious pessimism," constant quarreling and gloominess

Charles (d'Eyncourt)
1784–1861

"Inherited his father's instability and fretfulness"; spendthrift tendencies; expensive, grandiose activities and interest

Elizabeth Fytche
1781–1865

"Easygoing" and "sweet tempered"

George Clayton Tennyson
1778–1831

Vacillating moods "between frenzy and lethargy," spendthrift; alcoholic; "fits"; insanity

George
died in infancy, 1806

Charles
1808–1879

Addicted to laudanum, "complete nervous breakdown"; had to be segregated from outside world; extreme mood swings and "recurrent fits of psychopathic depression"

Mary
1810–1884

"...of a wild sort of countenance"; obsessed with spiritualism

Edward
1813–1890

Confined in insane asylum for almost 60 years; severe melancholia; death from manic exhaustion

Septimus
1815–1866

"Suffered from nervous depression"; frequent treatments for melancholia; "the most morbid of all the Tennysons"

Cecilla
1819–1899

"Mental disturbance and depression"; eccentric

Frederick
1807–1898

Irritability, eccentric, violent temper, and volatile; obsessed with spiritualism

Alfred
1809–1892

Recurrent depression that required treatment; trances, possibly epileptic, but not thought so by physician; possibly transient hypomanic episodes; "dwelling in an element of gloom"

Emily
1811–1889

Arthur
1814–1899

"Suffered much from depression"; one year in Crichton Institution for the Insane

Matilda
1816–1913

"Some mental derangement," occasionally attributed to childhood accident; religious obsessions; "did not entirely escape the black-bloodness of the Tennysons"

Horatio
1819–1899

"Strange personality was legendary"; "rather unused to this planet"; perceived himself vulnerable to the "weaknesses of the Tennysonian temperament"

(Bipolar) Manic-depressive illness   Recurrent depressive illness   Rage, unstable moods, and/or insanity

Spontaneous Stimulation A person in the throes of a psychotic episode is experiencing the results of the over-activation of brain circuits that ordinarily interpret sensory stimulation such as voices. But instead of starting from the ears and moving inward to the brain, the activation originates within the brain. "Schizophrenia involves a disturbance not within the whole brain but only in neural circuits that affect neurons in some of the most important regions of the brain governing thinking and perception," according to Jeffrey Lieberman, a psychiatrist and researcher at the University of North Carolina at Chapel Hill. Not only does a person with schizophrenia see and hear things that aren't there, but overstimulation in the frontal lobes makes it impossible for the person to focus and concentrate.

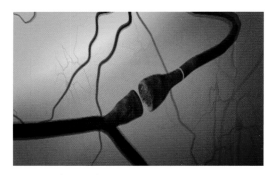

Random firing of certain key neuronal circuits may account for such hallucinatory symptoms of schizophrenia as hearing voices.

Myths and Misconceptions Over the centuries, explanations for schizophrenia have ranged from its being a disturbance in the bodily humors to demonic possession. Schizophrenics have been executed as witches and demons, confined in institutions, universally feared and shunned. More modern theories of the cause of schizophrenia have ranged from faulty parenting to in utero viral infections.

Insights into the true nature of the illness date from the 1950s. At that time a French neurosurgeon, Henri-Marie Laborit, was in search of a drug to calm patients before surgery. He found such a drug in promethazine, an antihistamine. But this drug acted differently from most antihistamines: It calmed the patients without putting them to sleep, even at high doses. Impressed with promethazine, the neurosurgeon obtained from a local pharmaceutical company a derivative of promethazine that worked even better. The drug, known as chlorpromazine (Thorazine) induced in Laborit's patients a state he referred to as "beatific quietude." Soon thereafter psychiatrists began using the drug to calm schizophrenic patients without over-sedating them.

By the mid-1950s, psychiatrists regularly employed chlorpromazine in the treatment of schizophrenia. But despite a stunning success with the drug, more than 20 years of subsequent research was needed before investigators discovered how the drug works.

Thorazine and other drugs like it—termed *neuroleptics*, from the Greek meaning "to grasp the neuron"—act by blocking receptors on neurons within the brain for the neurotransmitter dopamine. The drugs relieve the schizophrenic symptoms primarily by blocking a subclass of dopamine receptors known as D2

For reasons unknown, the sending neuron releases dopamine into the synapse between cells. Normally, the dopamine would be reabsorbed by the sending cell once the target cell is stimulated. Instead, there's a surge of dopamine that overstimulates the target cell. If this occurs in the frontal lobe, for example, it interferes with the ability to concentrate.

receptors, in the limbic (emotional) parts of the brain. A theory of the origin of schizophrenia emerged from the insight.

The Dopamine Theory Schizophrenia resulted, according to the dopamine theory, from an excess of dopamine at key receptors; treatment consisted of blocking those receptors with drugs like Thorazine. Researchers at the time believed that disturbances in dopamine levels alone could account for the illness. No one believes that now; at least not in the pure form of the dopamine theory. More likely, disturbances in dopamine balances are only part of a multi-neuronal chain that involves other neurotransmitters and brain chemicals. As a practical application of this new belief, the newest drugs for schizophrenia also affect other neurotransmitters such as serotonin and norepinephrine.

Unfortunately, none of the currently available drugs cure schizophrenia. While the drugs may decrease the incidence of delusions and hallucinations, other symptoms, such as apathy and withdrawal, remain largely unchanged. Some patients, like Sabrina Yeskel, are even more unfortunate: No drug so far has done away with her voices or the menacing figure in black who taunts her. This lack of response means not only further prolongation of Sabrina's agony but, in addition, a possible worsening of the damage done to her brain.

"We have emerging evidence that when a person goes on to develop persistent chronic symptoms, their brain suffers a loss of tissue involving the brain cells themselves and/or the many connections between the brain cells," according to Diane Perkins, a research psychiatrist at the University of North Carolina at Chapel Hill. "We believe psychosis is the most visible and obvious symptom of this toxic

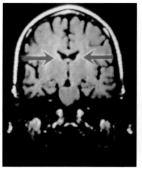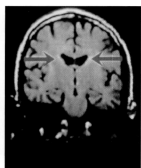

In MRI scans of a pair of identical twins, one shows the enlarged ventricles (far right) that characterize patients with schizophrenia. Although an identical twin of a schizophrenic has a 40-times greater chance of developing the disorder than a nonidentical twin, more than 50 percent of identical twins remain free of the disease. Environmental factors thus play an important—and as yet unknown—role in the development of this illness.

process. When we give medicines to stop the voices, we're stopping this toxic process. One theory is that the toxic process sets off a flood of neurotransmitters."

Glutamate's Role One neurotransmitter, glutamate, may play a particularly important role. At the time when neuroscientists first developed the dopamine theory, glutamate was not yet considered a key player in brain activity. Now neuroscientists have identified it as the most plentiful excitatory neurotransmitter in the brain. Excesses of glutamate are now known to be important contributors to stroke and other neurological illnesses. And it's also now considered likely that disturbances in the level of glutamate or other excitatory neurotransmitters may play an important role in schizophrenia.

Once again, the damage doesn't result from a glitch in only one neurotransmitter but rather from a disruption in the normal regulatory systems that keep the brain's many neurotransmitters in balance. "And the longer the illness goes on without treatment, the more damage it does to the brain; the more likely failures in cognition will occur; the more likely destructive brain changes will occur secondary to brain cell loss," according to Lynn Sikich.

Supporting the brain cell loss theory is the finding of increased ventricular size and reduced brain volume within medial temporal lobe structures in people with schizophrenia. Presumably, loss of neurons leads to a ballooning outward of the ventricles, in a process similar to the widening of a stream consequent to erosion along the banks of the stream. But is this ventricular widening present at the onset of the illness? Or is it a late-stage consequence of established schizophrenia?

CAT scan examinations of schizophrenics taken during the initial episode often show several abnormalities, suggesting that the brain alterations occur before the illness declares itself. Enlarged ventricles are among the abnormalities, along with a reduction in the volume of cortical gray matter and a loss of the normal pattern of cerebral asymmetry. (In normal people the right hemisphere is usually slightly larger than the left.) Moreover, since the normal pattern of cerebral asymmetry is established during the second half of fetal development, the failure of the schizophrenic brain to establish this asymmetry suggests the problem existed months before birth.

Abnormal Connections   One currently popular theory emphasizes the likelihood of abnormal connections between the frontal and temporal lobes. It's speculated that hallucinations may be a consequence of these faulty connections, since frontal–temporal lobe interactions are important in monitoring and recognizing "silent" speech as emanating internally rather than from the outside world. But whatever the ultimate causes and course of schizophrenia, why does it appear for the first time during adolescence?

According to Daniel Weinberger, "Adolescence is a very stressful time for the brain. There are chemical changes, hormonal changes, anatomical changes, changes in the expression of genes inside of cells. It's a time of great biological tumult. And the frontal lobe is fighting to adapt to the environment, to deal with all these inner instinctual urges. Indeed, it's difficult enough for people with a normal frontal lobe to make it through adolescence. But we believe that patients with schizophrenia don't have normal frontal lobes. We believe they didn't develop normally from early in life."

According to one speculation, "schizophrenia doesn't happen earlier because in younger kids those parts of the brain aren't that much in play," Lynn Sikich says. "As another way of putting it, schizophrenia disrupts the normal expected maturing of the brain."

In our discussions with neuroscientists we frequently heard reference to a "two-hit" hypothesis of the origin of schizophrenia. For the disease to develop, both "hits" must occur. The first hit is now believed to take place in utero or around the time of birth and to consist of altered brain development. In other cases, the hit may have consisted of an infection in the mother or infant that altered normal brain development. In either case, this first hit can frequently be managed without developing the illness until the time of the second hit, which can be abuse or trauma or, in the case of Sabrina, something as simple as separation from one's parents.

"Sabrina's psychotic symptoms first started when she was away from home for the first time," says her psychiatrist, Lynn Sikich. "She is clearly very attached to her family and depends on her family a lot. Thus the stress of that experience in the sleep-away camp may have provided the kind of second hit that pushed her into psychotic symptoms."

While Sabrina has not so far responded to any of the many drugs that her psychiatrist has tried, other schizophrenics are more fortunate.

Isaac's Story Isaac was a freshman in college when a strange series of events took place: "One night when I returned to my apartment, I began to hear some noises from the wall. That was when I first thought the apartment was bugged. When I told my roommates I thought there were speakers in the walls, they just said 'You're paranoid.' It was so frustrating because nobody would listen to me. Eventually I figured out the answer: I had a computer chip implanted in my head that acted like a cell phone. It would pick up signals, break them down, and then change them back into speech again. At one time the voices told me through the computer chip about a bomb in my car. When I looked at the back-seat of the car, I noticed that one of the strings in the upholstery was kind of undone. At that moment I was certain there was an explosive in there."

Isaac's mother has a slightly different recollection of events: "The first time I realized Isaac had a major problem was when he called to tell me he had a chip implanted in his head and that he was under surveillance. He became irate when I expressed reservations about this. Later I found out that a few days earlier, while gathered around a bonfire with some friends, a glass exploded and a small shard of glass had hit him in the face. And that experience seemed to have set off all this computer chip stuff. Later he became convinced there was a bomb in his car. That seemed real to him, very real. So he acted like there was a bomb. He did not realize that he was being controlled not by a computer chip but by his psychotic beliefs and behavior."

Today Isaac is out of the hospital and working at a restaurant as a short-order cook. He takes the drug Olanazopine (Zyprexa), a new-generation anti-psychotic that influences several neurotransmitters. "Zyprexa has brought me out of my own little world and into reality," says Isaac. "I'm reaching the final stages of recovery.

Isaac's symptoms of schizophrenia seem to have started after a shard from an exploding glass at a campfire hit him in the face. Researchers think the disorder may manifest if there is a developmental "first hit" that occurs in the womb, followed by some sort of traumatic incident in the vulnerable adolescent years.

I'm normal 99 percent of the time—actually more than 99 percent of the time. So I feel that's a great improvement."

According to Isaac's psychiatrist, Diane Perkins, the newer-generation anti-psychotic drugs like Zyprexa have conferred multiple benefits. "Not only have his psychotic symptoms improved but also his cognitive functioning, such as attention; memory; and his ability to organize, store, and use information. This is quite different from older antipsychotics that might have squelched the voices but left him still unable to learn, plan, and make choices. And since Isaac has shown improvement in all of these different areas, it's realistic for him to think about going back to school."

Treatment is only one of the goals of doctors wrestling with this devastating illness, however. Prediction and prevention are equally important. At the moment, no tests exist that provide reliable markers for the later development of schizophrenia. But there are certain factors that increase the likelihood of the disorder. Obstetric complications such as fetal distress, low birth weight, smaller head circumferences, and prolonged labor increase the risk for future schizophrenia. All of these complications are often associated with decreased blood and oxygen supply to the brain. This, in turn, frequently leads to bleeding in and around the ventricles—one possible mechanism for the ventricular enlargement consistently observed in schizophrenics.

If the brain abnormality responsible for schizophrenia begins in infancy and early childhood, might it be possible to detect hints and foreshadowings many years before the illness expresses itself?

"It's hard to imagine how a brain abnormality might be present at birth but not manifest itself in the individual's behavior or adjustment until so much later in

"Mom, I've seen this in Psych class ...."

When Courtney Cook was a senior in high school, he began to experience bizarre symptoms. He'd wake up in the morning and see things that weren't there: Gnats and sparkles swarmed into view as soon as he opened his eyes. He felt constantly "stupified," unable to focus on anything.

"All of a sudden I couldn't talk the way I used to be able to talk," he remembers. "I couldn't think the way I used to be able to think. I was responding to thoughts and ideas that weren't my own."

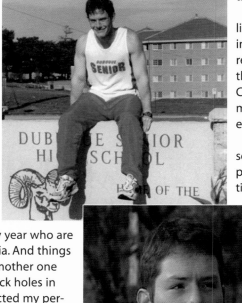

Like most teenagers, he didn't immediately tell anyone what was going on. But finally he went to his mother and said, "Mom, I've seen this in Psych class and I'm really terrified that I have schizophrenia."

Unfortunately, Courtney was right. He was one of the 300,000 Americans every year who are diagnosed with schizophrenia. And things got worse. He looked at his mother one day and told her she had black holes in her face. "Schizophrenia affected my perception of reality and twisted it," Courtney says. "My brain was flooded with things that I wasn't accustomed to."

Once an athlete and a good student, he found himself unable to concentrate. Like the hallucinatory gnats that were always somewhere in his field of vision, there were just too many things going on in his brain. Though he remained articulate and insightful, he had to fight to stay coherent.

"Everything just disintegrated," he says. "All the ideas I had about myself suddenly just disappeared.

After spending all those years with the same personality I was used to it; I liked it. And all of a sudden it was taken away."

Researchers believe the site of the worst damage in schizophrenia is the prefrontal cortex, the region that carries out the highest human functions of thinking, reasoning, and planning. But as Courtney discovered, the disorder also affects the brain's emotional centers.

MRI scans show that Courtney's limbic area responds poorly to images that usually elicit high response in normal people. "The thing that really crushed me," Courtney says, "was the lack of motivation, the flat affect, the emotionlessness."

Unable to feel either joy or sorrow, and unwilling to face the prospect of day upon day of emotional emptiness, he attempted suicide. "I didn't want to live the life I had been given," he said. "It was so contrary to what I wanted." Fortunately, his attempt failed.

Courtney has since been put on medications that seem to keep the worst of his symptoms in check. He is no longer suicidal, and he is even going to college.

Although he does daily battle with his disorder, he seems to have come to terms with it. "I'd rather have lived the life that I live now—as confusing and painful as it has been," he says, "than never to have been born at all."

life," according to Elaine Walker, a psychologist at Emory University, in Atlanta. "We directed our research therefore toward looking for very early signs of vulnerability—signs that indicated brain involvement prior to any behavioral expression."

Looking for Early Signs   Obviously, the most direct method of obtaining early developmental information would be to ask the parents. But few parents are trained observers and reporters of child development. Besides, even with their best efforts, parents often forget or misinterpret their child's early development in light of the child's subsequent schizophrenia. What researchers clearly needed was some way of *seeing* development rather than simply hearing about it.

"We asked the parents for home movies they had taken during their child's early years," Elaine Walker told us. "Home movies provided us with objective, unbiased information about how a schizophrenic developed as an infant and behaved as a child."

Walker observed delays in the future schizophrenic's motor skills such as crawling, walking, and handling small objects. She also observed spontaneous abnormal movements such as odd posturing of the hands and arms. Facial expressions also differed, with negative emotions occurring more frequently than positive ones. "Many of the children who went on to develop schizophrenia showed muted facial emotion: very few signs of excitement or glee but frequent facial expressions of sadness and distress."

Elaine Walker readily points out that none of the observed deviations provide unequivocal evidence of future schizophrenia. Many infants and children show "odd posturing" during the first year or so. In most cases—but not all—the posturing disappears. Even among those who persist in the movements beyond the third year, schizophrenia isn't an inevitable sequel. Nevertheless, after cautious and sensitive interpretation, the home movies have helped Elaine Walker arrive at a theory that helps explain why schizophrenia often takes decades to develop.

"Schizophrenia originates from a brain abnormality located beneath the cortex and involving the neurotransmitter dopamine. This abnormality remains relatively silent until adolescence. Those movement abnormalities, along with subtle reductions in positive emotion and increases in negative emotions, provide the only clue of what's to come. All of these expressions are linked with irregularities in dopamine transmission that, while present at birth, don't develop into identifiable illness until adolescence."

Walker's comments are supported by animal research. If lesions are made in the hippocampus of newly born rats, nothing much happens until young adulthood when the animals become supersensitive to stress and exhibit behavioral abnormalities that are worsened by drugs that activate the dopamine system.

In humans, it's considered likely that the brain lesions responsible for schizophrenia remain silent until the normal brain maturation in adolescence leads to the use of brain circuits not greatly developed in children.

"During adolescence, schizophrenics first start to experience problems getting along with others, anxiety in social situations, depression, and unusual thoughts and beliefs. Hormones may play a role here, since adolescence is marked by an increase in stress hormones. While this increase is perfectly normal in adolescence, it seems to be a problem for those adolescents at risk for the illness," according to Walker.

Fortunately, most of us survive and overcome the challenges of adolescence. This is largely due to the gradual maturation of our prefrontal cortex. As this pivotal brain area matures, we become less driven by impulse, more likely to weigh consequences, less likely to restrict our mental horizon to momentary rewards. This doesn't imply a rigid demarcation, incidentally. We are all familiar with adults who remain immersed in adolescent conflicts: rebelliousness, impulsiveness, drug and alcohol abuse. The life trajectories of these tragic individuals illustrate an important point: Adolescence doesn't involve clear borders and sudden ends but exists along a continuum.

Leaving Childhood Behind

With each passing year the adolescent leaves childhood farther behind and inches closer to the world of adult privileges and responsibilities. This isn't a sudden shift but a gradual one. It's no wonder, therefore, that adolescents often experience identity problems. The adolescent dwells in a no-man's land: no longer a child but not yet an adult.

As we have seen, adolescence is both biological and a state of mind. If the adolescent's brain develops normally, he or she will reap the benefits of plasticity: normal emotional-cognitive development. But interference with that plasticity by diseases or drugs can lead to emotional and cognitive disturbances that may last a lifetime. Normal transition from adolescence to adulthood occurs in concert with enhancement in the functioning of the frontal lobes.

Many of the issues of adolescence continue beyond the teen-age years. None of these issues are more important than the integration of thoughts and feelings. Indeed, the strengthening of this capacity to balance emotion with the dictates of reason is the distinguishing feature of the transition from adolescence to the next stage of brain development: adulthood.

TO THINK BY FEELING

THE ADULT BRAIN

Adulthood is supposed to correspond to the period in our lives when reason rather than emotion rules. While children and adolescents may be forgiven if they allow feelings to determine behavior, adults are required to act reasonably. So strong is this emphasis on reason that our entire legal system is based on it. Judges and juries perennially search for an answer to, 'What would the reasonable man or woman do under such circumstances?'

But a moment's reflection on our own and other people's behavior confirms that few of us are always as reasonable as we would wish others to believe. We may become cantankerous whenever certain subjects are introduced into conversation, or "fly off the handle" when we can't get our way. Most interesting of all, many of the things that we believe "stand to reason" are often based on assumptions and experiences that have more to do with emotions than reason. We're likely to differ in our emotional reactions depending on such things as our place of birth (a small town versus a large city), our nationality, our religious upbringing (or lack of it), the size of our family, even the political affiliation of our parents. So strong is the influence of our emotions on our reasoning that the great nineteenth-century psychologist William James devoted an entire essay to the subject, aptly entitled "The Sentiment of Rationality." Indeed, reason and emotion are as intertwined as the threads in an oriental carpet.

An example from my own experience illustrates the point. During a recent thunderstorm I was awakened late at night by the ringing of the phone next to my bed. Since I'm a practicing neurologist, late-night calls aren't that unusual. But this call was different. I could hear somebody laughing in the background. After being aroused from a sound sleep, I wasn't in the mood for levity. A male voice asked me, "Do you own a bearded collie named Bobbie?" I tried to conceal my annoyance at this stranger who had awakened me to make a frivolous inquiry about my dog. Less than an hour earlier I had observed Bobbie nervously pacing around the basement in response to the thunder. As far as I was concerned, the dog was still there. "Yeah, I own a beardie named Bobbie. Why are you calling to ask me a question like that at one o'clock in the morning?" More laughter from the stranger triggered more annoyance on my part. "Because I've got Bobbie here in my SUV. He ran out in front of me and I nearly ran him over. I got your phone number from his collar, and if you give me your address I'll bring him over." Annoyance quickly changed to relief and joy that Bobbie was unhurt. But an instant later I experienced apprehension. Bobbie couldn't have gotten out of the house on his own. Who was this person? Could this be some scheme to gain entrance to my house? "Hold just a moment," I requested, and I ran to the basement to discover Bobbie gone. Fifteen minutes later the man and his wife returned Bobbie to his grateful owner. Later I learned that one of my daughters had inadvertently let Bobbie out of the house when returning home through the basement entrance.

The Gamut of Emotion   In the space of less than a minute or so I had experienced annoyance, hostility, apprehension, fear, relief, elation, embarrassment (at my annoyed tone on the phone), a brief period of suspicion, and finally, deep gratitude. In short, my brain didn't respond to learning about Bobbie's narrow escape simply on the level of "information." Instead a roller coaster of emotions accompanied each new twist in the story.

Most people can readily bring to mind episodes in their own lives where reason and rationality shared the mental stage with emotions and feelings. We always react emotionally to even the most trivial of happenings. Imagine yourself coming home to a phone message on your answering machine from a former

Research suggests that each of our feelings of emotion has a specific neural circuitry, including some brain regions not usually associated with emotion, according to the University of Iowa's Antonio Damasio. In his studies of volunteers remembering emotionally powerful personal events, reliving a specific emotion caused significant changes in certain parts of the brain, as seen in the PET scans below. Red hues indicate areas of increased activity; purple hues indicate regions of decreased activity.

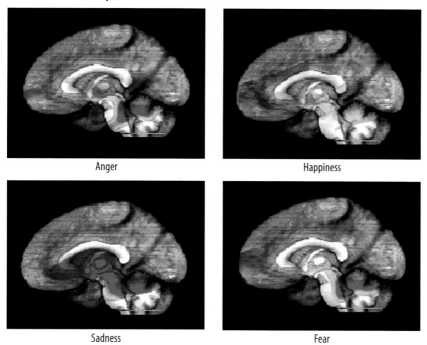

Anger　　　　　　　　　　　　Happiness

Sadness　　　　　　　　　　　　Fear

"intimate" you haven't heard from in a long time. What could be the problem? Is there a problem? Why this call? Why now?

Thanks to contemporary neuroscience, we can now begin to understand what's happening in our brain when we experience certain emotions. It's early in the game, and few emotions can be pinpointed, but we're edging closer to a scientific understanding of our emotions. As a starter, neuroscientists are confirming our subjective insights that the adult brain is not neatly divisible into areas exclusively committed to either reason or emotion. Neuroscientists learned this by studying the fear response in animals.

Although fear may sound like an odd emotion to select for study (why not happiness?), fear gets the nod because it can be observed (or at least inferred) in all animals. The universality of fear is based on its survival value: It prepares an animal to defend itself against potential harm. Fear, therefore, is the emotion neuroscientists have studied most intensively.

Source of Emotions

English speakers may choose from an array of overlapping words to describe their subjective experience of emotions and feelings, and researchers have spent decades trying to define them (see below). The current consensus seems to be that our emotions result from a complex interplay of biology, behavior, and cognition. Emotions evolved to help us survive both as individuals and as a species. Without emotions to guide us, we would be incapable of either decisions or plans.

Research suggests not one emotional system in the brain, but many. After all, the emotions involved with defending against danger differ from those involved with finding food and mates or with caring for offspring. Each may well involve different brain systems that change for different reasons.

Shown at right are some of the brain regions that play a role in emotions. This loose assembly of structures and pathways, known as the limbic system, evolved very early in brain development. Nonetheless, extensive interconnections exist between the limbic system and the rest of the brain. New research is showing that areas of the more recently evolved cerebral cortex are critical to our ability to shift from instinctive reaction to more considered action—that is, to assess the current emotional situation and decide what action we should take next.

In a system devised by psychologist Robert Plutchik, emotions are analogous to the colors on a color wheel. Eight primary emotions are arranged as four pairs of opposites, with the most intense emotions in the inner circle of the exploded view, corresponding to the top of the cone. The emotions listed in the blank spaces of the exploded view are what Plutchik calls the "primary dyads"— emotions that are mixtures of two of the primary emotions. So, for example, awe is a mixture of fear and surprise; a mix of fear and trust produces submission; and a mix of trust and joy produces love.

The areas of the brain depicted below mediate most of the basic human drives and emotions—as well as the instinctive, or automatic, behaviors important to our survival. A key player in our emotions is the tiny, almond-shaped amygdala, which receives information directly and indirectly from many parts of the brain. For example, the olfactory bulbs deliver their input directly to the amygdala and to the olfactory cortex (on the inner part of the temporal lobe), explaining why the sense of smell is a powerful influence on our emotional behavior.

The thalamus relays information from other senses to the amygdala and to the parts of the cortex that interpret vision, hearing, taste, and touch. Those regions then send their signals to the amygdala. The amygdala,

for its part, sends signals to the hypothalamus, which, along with the pituitary, controls autonomic nervous system functions, such as breathing and heart rate.

The hippocampus, a critical structure in memory, gets input from the amygdala and exchanges signals with the entire cerebral cortex. The fornix and the septum are both involved in routing neuronal signals between the hippocampus and the hypothalamus, as does the mammillary body. The cingulate gyrus acts as a relay station between the cortex and structures such as the hypothalamus and the hippocampus.

In most people, the prefrontal cortex modulates the responsiveness of the amygdala. Usually, when prefrontal cortex activity is high, amygdala activity is low, and vice versa.

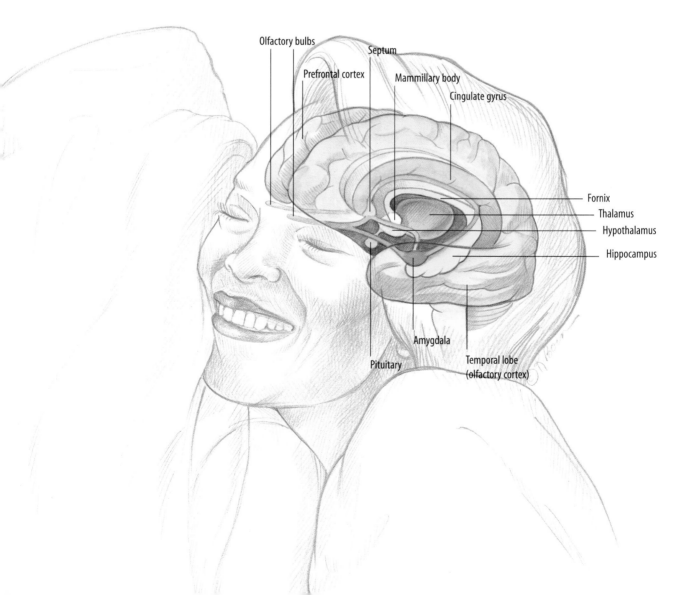

As you probably recall from your psychology classes, Pavlov first demonstrated at the turn of the twentieth century that dogs could be conditioned to salivate at the sound of a bell. All that was required was a previous session when the sound of the bell occurred while the dog held a tasty morsel of meat in its mouth. Pavlov suggested that the association of the bell and the meat created a connection in the brain so that the sound could serve as a substitute for the meat and thereby elicit salivation.

The Fear Response   In the laboratory, fear conditioning typically involves exposing a rat or other animal to the combination of an electric shock with a sound. After a few shocks the animal responds to the sound alone, which causes the rat to "freeze" in its tracks for a few seconds. The animal's blood pressure and breathing rate also increase as it prepares itself to respond. A similar fear response occurs in humans.

Imagine yourself at a public sporting event. As you're making your way through the crowd, you suddenly hear a loud explosion. You momentarily "freeze" as you rapidly compute various scenarios. Was that the sound of a terrorist bomb, a teenage prank, or a misfire from a car just outside the entrance of the stadium? And if it was a bomb, did it go off anywhere near where your friends are sitting? After a few seconds of stunned immobility during which your blood pressure and respiration increase, you finally realize to your horror and alarm that the sound resulted from the explosion of a bomb. Furthermore, you recognize that you're very close to the site of the explosion. With this recognition, and out of fear that another bomb may be about to explode, your immobility gives way to panic as you join the crowd in a mad race away from the scene.

The scenario just described actually happened at the Olympics in Atlanta in 1996. A video captured the two-part response of those near the explosion: an initial short response marked by freezing in place, followed by a second response of panicked flight.

Despite the vast chasm separating the laboratory rats from the fans in Atlanta, the responses were similar. Indeed, this similarity in response suggests to Joseph LeDoux, a neuroscientist at the Center for Neural Science at New

York University, that similar brain mechanisms involving the amygdala are at work in the fear response.

Encoding Emotional Memory   Earlier we discussed the amygdala as part of the pleasure circuit. This small region, named for its almond shape (*amygdala* is from the Latin word for almond), is important for encoding the emotional aspects of memory. A memory without any particular emotional overlay, in contrast, is encoded principally by the hippocampus, the seahorse-shaped structure connected to the amygdala. But if that emotionally neutral memory takes on emotional significance, the amygdala comes into play. Can you remember where you were and how you felt when you heard about the Oklahoma City bombing or the death of Princess Diana? If you can, it's because the emotional impact of the tragic event activated the amygdala and other emotional areas in your brain to create what psychologists call a "flashbulb memory."

But as LeDoux and others discovered, the amygdala is not just important in memory but also plays an important role in generating fear. Electrical stimulation of the amygdala induces every measure of conditioned fear, including freezing, stress hormone release, blood pressure elevation, and a general firing up of the autonomic nervous system responsible for the "fight-or-flight" response. Destruction of the amygdala, in contrast, leads to the abolition of all of these components of the conditioned fear response.

"We can think of the amygdala as the hub of a wheel; all of the spokes coming into it are different pathways from different parts of the brain," LeDoux says. "The amygdala is, in essence, involved in the appraisal of emotional meaning and is the site where trigger stimuli do their triggering. Any emotionally arousing situation, especially fear, activates the amygdala to stimulate the hypothalamus, leading to the release of hormones and other chemicals. It's all part of the process of detecting danger and mobilizing a response. And since the amygdala activates the hippocampus and cortical areas, you will retain an embellished, stronger, more enduring memory of the dangerous situation."

Additional experiments in LeDoux's laboratory revealed the existence of a "high" and a "low" road for the conditioned fear response. The direct pathway from the thalamus to the amygdala is shorter and faster. As a result of activation of this

Evolution has designed the brain so that a frightening stimulus—hearing an explosion, say, or falling off a motorcycle—is picked up by the thalamus and sent directly to the amygdala. Although different parts of the amygdala receive input from different senses, they all communicate with the central nucleus, which, in turn, communicates with the brain stem to set in motion the body's "fight-or-flight" responses. Experiences that trigger the amygdala can lay down such strong memories that, in some people, a similar stimulus even years later will set off a panic attack—a symptom of post-traumatic stress disorder.

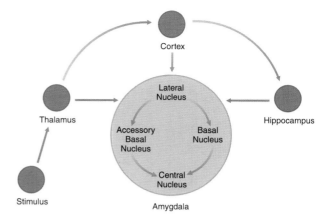

pathway, you start and freeze at a loud sound before you know the source of the sound. The second pathway—from the thalamus to the cortex and then to the amygdala—allows additional time for the cortex to come up with an explanation for the sound.

What is the usefulness of such an arrangement? In a dangerous situation, that extra time provided by the direct thalamus-to-amygdala pathway might mean the difference between life and death. Was that sound a bomb? Better to freeze for a second rather than continue on your way and be injured or killed by a second blast. Or suppose that during a walk through the woods you detect a curved shape on the ground. Rather than continuing forward, you're better off freezing in place for the interval required for your cortex to work out whether you've encountered a snake or just a curved stick. If it's a stick, that initial arousal by the amygdala and the consequent fear response won't cost you more than a few wasted seconds from your power walk. But if the object turns out to be a snake, that momentary pause might save your life during those few seconds while the cerebral cortex decides what's in front of you and how best to respond.

Consciousness Takes a Back Seat   Notice that consciousness plays no role in the fear generated by activation of the amygdala. Your initial response to that loud explosion or to that curved object in the grass occurred several seconds before you consciously registered it. Only then did you attempt to decide on the proper response.

According to LeDoux, "This difference between the processing of the high road and the low road can result in people having emotional reactions they don't understand. Furthermore, amygdalae are wired up differently, based on the genetic instructions and different experiences that shape synapses as they develop

Saved by the "Low Road"

In an emergency, the instinctual reaction of the "low road" can mean the difference between a nasty accident and a near miss. Two motorcyclists go down in the middle of a race, their bikes useless hunks of metal as the pack swarms around them. In their brains, the amygdala is instantly in high gear, sending signals that trigger a variety of physiological responses in the brain and body.

As instinct and the body's autonomic nervous system take over, less urgent functions are suppressed. Abdominal organs, face, and genitals donate blood to the brain, heart, and leg muscles, fueling them with neurotransmitters, oxygen, and blood sugar. Hungry for sugar, the brain activates liver and fat cells to release glucose and fatty acids for energy. Insulin production slows down to give the brain even more glucose.

Lung muscles relax, allowing easier airflow. Heart rate and blood pressure rise, prompting the brain to send cooling signals to the body. Water is expelled through the palms of the hands, the armpits, and the face, resulting in a clammy "cold sweat." Meanwhile, the adrenal glands are pumping out high-octane adrenaline into the bloodstream.

One rider's systems function flawlessly. As the other rider stumbles and falls back to the track, he sprints out of harm's way.

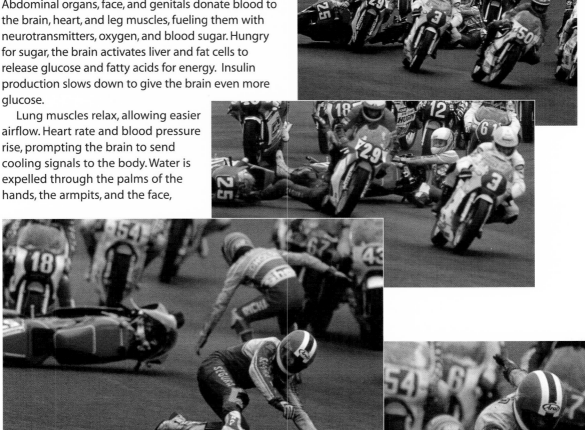

during childhood, adolescence, and adulthood. Depending on your genes and your experiences, you may have an amygdala that's more or less sensitive, more or less reactive to the same experience."

In most people the prefrontal cortex modulates the responsiveness of their amygdala. Imaging studies reveal that when the prefrontal cortex activity is high, amygdala activity is low, and vice versa.

According to James McGaugh, founding director of the Department of Psycho-biology at the University of California at Irvine, "You've got a clash between the prefrontal cortex, which is trying to make sense of what's going on at the moment, and the amygdala, which instructs the brain to induce emotional arousal."

Unfortunately, the amygdala too often trumps the prefrontal cortex in almost all of us. "Fear is the only emotion that overcomes greed," as one rather cynical but prescient stockbroker commented to me during a sudden sell-off in the market. Historically, Charles Darwin once put the preeminence of fear to a test:

"I put my face close to the thick glass plate in front of a puff adder in the Zoological Gardens, with the firm determination of not starting back if the snake struck at me; but as soon as the blow was struck, my resolution went for nothing, and I jumped a yard or two backwards with astonishing rapidity. My will and reason were powerless against the imagination of a danger which had never been experienced."

All of us can point to personal experiences that confirm Darwin's test of his ability to inhibit automatic fear responses. When gripped by inappropriate fear we use our prefrontal cortex to talk ourselves out of it: "That sound I just heard was a creaking in the staircase and not an intruder." But when that inhibition doesn't take place, the amygdala–prefrontal cortical relationship flip-flops: The more basic primitive responses of the amygdala are released, and fear becomes uncontrolled and unregulated.

The Fear Response Gone Awry   Post-traumatic stress disorder (PTSD) is the most dramatic example of the fear response gone awry. PTSD research is helping neuroscientists unravel the fine threads of that oriental carpet in which reason and emotion are so intertwined.

The official definition taken from criteria established by the American Psychiatric Association includes four main components:

- A terrifying or horrifying experience that threatens life or inflicts serious injury to oneself or others.
- Reexperiencing of the original trauma via recurrent, intrusive, distressing thoughts, images, dreams, or flashbacks.
- Attempts to avoid activities or people associated with the trauma, often accompanied by feelings of detachment, loss of interest, and diminished pleasure.
- Arousal symptoms such as insomnia, difficulty concentrating, an exaggerated startle response, and a generalized hypervigilance that detects threats where no real threats exist.

Of these four components, flashbacks are the most distressing for the PTSD sufferer. He or she experiences the traumatic experience as if it's happening in the present moment rather than at some point in the past. A rape, a car accident, a mugging—whatever the trauma, it seems to be happening *now*.

Lost in Time In addition to a strong sense of reliving the past, flashbacks involve a momentary loss of context involving time and place. When in the midst of a flashback, the afflicted person loses track of surroundings and, under extreme circumstances, may actually behave as if back in the original traumatic situation. Such an utter and complete absorption in an event from the past seems unbelievable unless one has personally observed such a case.

For instance, a patient of mine was operating a subway train on the morning a woman jumped to her death in front of the train. As my patient told me about what she had seen before and shortly after the suicide, her voice rose and cracked, her eyes dilated, her breathing rate increased. At that moment, she seemed to be looking through me or beyond me, as if fixing on something in the room I couldn't see. When I asked what she was experiencing, she reoriented, started to cry, and said, "I just can't get the sight of all that blood out of my mind."

Not surprisingly, neurologists and psychiatrists have devoted much time and effort toward understanding the brain processes underlying PTSD caused by severe stress. But before they come up with a satisfactory explanation for PTSD, they must understand what happens in the brain under conditions of mild stress.

Everyday Stress   Take everyday stress, for instance. I'm not referring here to stresses like that of my patient, which are sufficiently outside of everyday experience to leave in their wake crippling and overpowering mental disabilities. Instead, I'm referring to garden variety stresses we all encounter. Public speaking is a notable example. According to several surveys, most Americans list public speaking as one of the most dreaded and stressful of life's varied experiences. Because of this near universal fear associated with it, public speaking makes an excellent vehicle for learning about stress.

In his Emory University laboratory, in Atlanta, Charles Nemeroff measures stress responses in volunteers before and after they give a 10-minute speech to three members of his staff. When finished speaking, the volunteers then carry out a second test most people also find mildly stressful: mentally subtracting a two-digit number from a four-digit number, such as, how much is 4,000 minus 17? Nemeroff finds that the level of stress hormones—principally cortisol, secreted by the tiny adrenal glands, which sit atop the kidneys—becomes much higher than normal in the blood of people with stress-related psychiatric disorders.

Cortisol is one of the more interesting and significant actors in our emotional life: Increased secretion is well known to be associated with depression. The more severe the depression, the greater the magnitude of adrenal steroid hypersecretion. Depression isn't believed to result from increased cortisol per se, however, but rather from the influence of a chemical within the brain, corticotropin-releasing factor (CRF, for short), that controls cortisol secretion. CRF not only influences cortisol levels but also the levels of norepinephrine, a second stress-related hormone.

Nemeroff calls CRF the "brain's stress hormone." When injected into the brains of laboratory animals, CRF produces many of the symptoms of depression: loss of appetite, sleep disruption, decreased sexual drive, and a phobia related to novelty or new environments. On the basis of this finding, Nemeroff suggested the CRF hypothesis of depression: Depressed patients hypersecrete CRF, which acts on the pituitary to release adrenocorticotropic hormone (ACTH), which

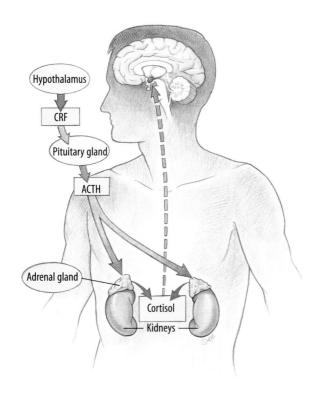

The Stress Response. An important phase of the body's response to stressful stimuli takes place along the so-called HPA (hypothalamic-pituitary-adrenal) axis. The hypothalamus produces the hormone CRF, which stimulates the pituitary to release ACTH. This hormone in turn stimulates the adrenal glands, which sit atop the kidneys, to release the stress hormone cortisol. When we're reacting to a fight-or-flight situation, cortisol increases the body's energy supply. It also acts in a complex feedback loop (dotted arrow) to the hypothalamus to regulate production of CRF. If the HPA axis is chronically activated, the overproduction of stress hormones can produce illness and depression.

then travels through the blood to the adrenal gland and induces it to pump out excessive cortisol.

High CRF Levels   In an initial test of his CRF hypothesis of depression in 1984, Nemeroff looked at the spinal fluid of depressed patients from Hungary— a country with one of the highest suicide rates in the world. The CRF levels were two times normal. At autopsy of depressed patients who had committed suicide, the number of CRF-producing cells in the victim's hypothalamus was two and a half times normal. As he extended his studies of depression and the CRF-depression link, Nemeroff came up with an unexpected, even shocking finding: Depressed women with a history of neglect or early physical or sexual abuse in childhood or adolescence showed an increased response to stress when compared to women without such a history. This included higher levels of cortisol and increased spinal fluid concentrations of CRF.

"We developed the hypothesis that if you were traumatized early in life, provided you had the genetic predisposition, you would have some type of plasticity-associated increase in your CRF neuronal activity that resulted in CRF hypersecretion," Nemeroff told us during our interview in his laboratory.

Certainly such a predisposition can be demonstrated in animals. In a "rat model of childhood neglect," Nemeroff separated pups from their mothers for three hours daily between days 12 and 21. In rats such a period of separation leads to neglect resulting from the mother rat's loss of interest in the pup when it is returned to her.

After their return to the cage and the ensuing neglect by the mother, the rat pups were stress tested at 90 days (the equivalent of young adult age). In the brain, spinal fluid, and blood of the rats, Nemeroff found all of the markers of CRF hypersecretion. In essence, early abuse or neglect led to a lifetime vulnerability to stress. Would the same rule hold for humans?

A Lifetime Vulnerability To find out, Nemeroff started administering stress tests—the public-speaking and number-subtraction exercises—to adult women with and without a history of abuse or neglect. He has found elevations in the cortisol and ACTH levels in response to these mild stressors in those women with backgrounds of abuse or neglect as children. In the controls (women of the same age, background, and experience except for childhood abuse or neglect) the levels rise slightly and return quickly to normal. But in women who were subjected to abuse or neglect as children, the levels escalate dramatically. If the women are currently depressed, the levels are even higher.

"Their response to the mild stress of giving a short lecture and then performing a simple arithmetic calculation is viewed as the ultimate stressor," according to Nemeroff. "Their adrenal glands couldn't secrete any more cortisol. This study is the first to demonstrate that early life trauma in women results in a permanently hyperactive stress response. Even though the traumatic events occurred approximately 20 years ago, their stress response is now permanently hyperactive."

Nemeroff's stress research illustrates an important point: The adult brain cannot be considered in isolation from the brain of the child and adolescent. Deviations from what pediatricians refer to as the "normal expected environment" can induce permanent changes in the brain's growth, structure, and functioning.

Thanks to the brain's plasticity, experience at every period of the life cycle exerts an influence on the adult brain that, given the appropriate circumstances and testing, can be revealed decades later. Moreover, these influences can leave permanent scars.

Women who have been neglected or abused as children are more likely to become depressed, anxious, or abuse drugs. Such considerations are particularly relevant in light of the increasing incidence of child abuse. And as Nemeroff's findings illustrate, the immediate effect of abuse is only the tip of a very ugly iceberg.

Plasticity, as we have observed in earlier chapters, has both positive and negative connotations. Love, nurturing, and support allow the brain to follow a normal path of development; abuse and neglect stunt the brain's progress and produce permanent defects.

Sometimes—as with women abused as children—the defects may be subtle and require special investigations like Nemeroff's public-speaking test to bring them to light. But the defects are there, nonetheless, and with tragic consequences not only for the victim of the early abuse but for their spouses and children. For instance, the incidence of divorce and instability is higher among women with histories of abuse or neglect. Along with an increased incidence of depression, these women are also more likely to develop borderline personalities marked by emotional instability, frequent mood swings, rages, impulsive sex, suicide attempts, and other self-defeating and self-destructive behaviors. And, as mentioned earlier, PTSD is also more prevalent among both men and women who have suffered abuse as children.

Measuring Vulnerability   While Nemeroff's findings provide a foundation for understanding the destructive effects brought on by the more severe stresses associated with PTSD, researchers need some way of directly measuring a person's *vulnerability* to the illness. Roger Pitman, a psychiatrist and specialist in the treatment of PTSD at Massachusetts General Hospital, has developed such a test.

First, Pitman requests the person to describe his or her traumatic event in vivid detail. "Run it through your mind and try to relive it just as if it's happening now," he tells the subject. Afterwards, Pitman composes a 30-second narrative of the event. He then takes the person into a laboratory and plays a tape recording of the narrative while measuring changes in his subject's heart rate, muscle tension, and sweat gland activity.

Johnny Cortez

Here is how Johnny Cortez remembers the auto accident that set off his PTSD: "It was 6 AM on a normal day and I was headed to work. I stopped at the red light at the intersection. As the light turned green, I started into the intersection. I heard a horn, and I took a look to my right side. That's all I can remember." A year later the arm and leg fractures Johnny sustained in the accident have healed, but he is still unable to free himself from tormenting memories and crippling fear: "I'm doing my best, but I'm having a hard time forgetting the accident. If somebody slams a door or makes a sudden noise, I jump and my heart starts beating fast. When I'm driving I keep rechecking to make sure the light is really green. I begin to think 'Maybe I'm going to crash; maybe somebody is going to hit me.' I have nightmares, but when I wake up I can't remember them. I spend hours trying to get to sleep but just when I'm drifting off any kind of noise will wake me back up and I lose control."

At the VA Research Center in Manchester, New Hampshire, Johnny is a volunteer in a study that monitors his level of anxiety. Roger Pitman and his team measure bodily responses such as heart rate and sweating as Johnny sits listening to a narrative of his accident gleaned from his own description of what happened (middle and far right). As the narrative proceeds, Johnny's heart rate speeds up and his hands get sweatier. He is reliving what happened that morning.

The goal of treatment is to modulate the activity of the overactive amygdala by enhancing the

Persons with PTSD—in contrast to subjects who don't develop the disorder after a traumatic event—show greater bodily responses to the narrative. Their heart rate increases, they start sweating, their muscle tension increases—responses more appropriate to the original event than to the comparative tranquility of a laboratory. In addition, the PTSD subjects report feeling greater anger or sadness.

Reliving, Not Just Remembering   "These responses indicate that PTSD patients are reliving and reexperiencing in the laboratory the events that provoked their PTSD. While those people who don't have PTSD are *remembering* the events, those with PTSD are *reliving* them," says Pitman.

Joseph LeDoux's research on conditioned responses provides an explanation for Pitman's findings. First, the traumatic event (the unconditioned stimulus) induces intense emotions such as fear, helplessness, or horror (unconditioned response). At the same time, environmental cues present at the time of the traumatic event (conditioned stimuli) become mentally associated with the experience. Subsequently these conditioned stimuli invoke intense emotional responses (conditioned responses), even under non-threatening conditions.

On the basis of LeDoux's work, Pitman suspected an important role for the amygdala in the establishment of the conditioned reflex responsible for PTSD. Starting with the activation of the fear response by the amygdala, the message is relayed to the hypothalamus, which orchestrates the bodily accompaniments of

restraining influence on the amygdala of the frontal lobes. If the treatment works, Johnny will find each "reliving" of the accident less traumatizing, less physiologically arousing.

Another approach, although not applicable to Johnny, is to administer drugs soon after an accident that block the action of adrenaline—the "fight-or-flight" hormone. As a result, the memory for the accident would be encoded minus the adrenaline-mediated emotional arousal. "So even if the person relives the situation, even if they have flashbacks, they won't have the adrenaline making too strong a memory," says James McGaugh of the University of California at Irvine.

fear (rapid heart rate, startle, enhanced sweating). Rather than making a onetime response, the amygdala reactivates each time the person encounters some component of the original experience.

"Although my findings and the animal experiments provided a persuasive argument for what takes place in PTSD, the evidence was still indirect," says Pitman. "We needed to look at the amygdala directly. And to do that we had to turn to imagery."

Employing PET and functional MRI recordings, Pitman repeated his experiments using the tape-recorded descriptions of traumatic events. As people with PTSD relived their trauma, Pitman found the amygdala was one of the key areas where blood flow increased. "So now we had some direct evidence for the involvement of the amygdala in the learned fear response, the fear memory."

The next step was to explore via brain imaging the mechanisms involved in the establishment of the fear response in PTSD. Thanks to LeDoux's work, Pitman already knew the answer in rats: activation of the amygdala via the low road from the thalamus. But does the same process occur in humans? Certainly comparable experiments in humans—combinations of isolation, electric shocks, and loud noises—would be unethical, as would any attempt to deliberately create a traumatic experience. So how could he discover the basis for the amygdala's increased activity in PTSD?

Unconscious Reaction to a Face   A clue emerged from some earlier research on normal volunteers asked to scan a series of pictures of human faces. A fearful face increased activity in the amygdala in fMRI and PET scans. Even more intriguing, the increase occurred before the subjects consciously processed the facial expression as fear inducing. This finding, of course, was very suggestive of activation of the amygdala via the direct pathway from thalamus to amygdala.

Curious as to how they would respond, Pitman showed the pictures to people with PTSD. He found a larger amygdalar response to fearful facial expressions, larger by far than for people subject to trauma who had not developed PTSD (their response was the same as that of people who hadn't experienced any trauma).

"So here was what we needed: a convincing demonstration that the amygdala is more reactive in PTSD. We had suspected this, but now we had the experimental evidence to prove it."

But this finding of increased amygdalar responsiveness in the person afflicted with PTSD doesn't answer a vital question: Which came first? Are people who develop PTSD born with an overly sensitive amygdala and therefore a predisposition to the disorder? Or does the traumatic experience sensitize the amygdala and, as a result, the person goes on to develop the disorder? We encounter here, as we have at other times in our exploration of the brain, the classic nature-nurture question: Are we dealing with genes or environment?

Twin Studies   To explore these questions, Pitman turned to the gold standard in distinguishing nature from nurture: twin studies. He looked at men with PTSD resulting from combat in Vietnam who had identical twin brothers who hadn't gone to Vietnam, although most of them were in the service. Of 150 such twin pairs, about a third developed PTSD, whereas the remaining two-thirds did not. In Pitman's experiment, each participant heard a sudden, loud noise. As expected, the responses were greater among the men with PTSD compared to the twin without PTSD.

"But then we came up with the really interesting finding. We tested the twins of the men with PTSD and found that they didn't overrespond to the loud noise. That's not the finding you would expect if the overactive amygdala—as measured by the overreactive response to the noise—had been inherited. If inheritance were the critical determinant, you would expect that the identical twin who shares the same genes would respond the same as his twin brother. So we can conclude that it was a life event, something that happened in a particular environment, that changed the reactions of the person with PTSD."

As Pitman talked about his research we put a question to him that was bothering all of us: Is everybody at risk for PTSD?

"We know that anybody who experiences a sufficiently frightening or disturbing event can develop PTSD. Nobody is immune. But we also know that some people are at greater risk than usual for developing PTSD after a traumatic event." Who are these people? One of the most important risk factors for PTSD is a history of prior traumatic experiences. People who have such a history are more likely to get PTSD after military combat, for example. Their sensitized amygdalae lead to stronger emotional reactions as a result of increases in stress hormones.

To James McGaugh, PTSD responses represent both "ordinary emotion gone haywire" and a hormone-driven disturbance in memory. "Take, for example, a car accident, mugging, or other stressful experience. You form a memory for the event. Each time you think about that event, rehearse it in your memory, the same emotional reaction is experienced again. In addition, the same hormones that etched that memory are released again," says McGaugh. "The next day you have another flashback and another inadvertent, unplanned rehearsal of the stressful experience. Each rehearsal of that experience, and the accompanying hormonal release, causes that memory to grow stronger. Left unchecked, this overremembrance leads to nightmares and more flashbacks. Eventually the disorder takes over your life."

The Nonverbal Low Road   Not surprisingly, treatment of PTSD is demanding, elusive, and rarely successful with psychotherapy alone. That's because the "quick-and-dirty" low road (as Joseph LeDoux refers to the direct thalamus-to-amygdala pathway) doesn't deal in words or concepts. Once a conditioned fear response is established, the person with PTSD reexperiences panic, fear, and total behavioral disorganization when faced with the same circumstances associated with the original conditioning stimulus (e.g., the street corner where a mugging took place.) While psychotherapy can modify the "high road" so that the person understands that his fear of a street corner is irrational, it doesn't modify the low road and the resulting flood of reason-overwhelming chemicals released via the autonomic nervous system.

In fact, for the person afflicted with PTSD, a potential for danger does not even have to exist. Even an innocuous sound can arouse a fear so powerful that it overcomes any attempts at persuasion, reasoning, or exhortation. As Roger Pitman's work suggests, this triumph of fear over reason may result from an amygdala that is oversensitive or hyperresponsive compared to the amygdala of the average person. But does this alteration in normal functioning by the amygdala mean that PTSD is incurable? Not at all.

Modifying the Fear Response   While the amygdala plays a pivotal role in determining emotional responses such as fear, other brain areas can modify and sometimes even inhibit the fear response. Plasticity enables the brain to fight back. Roger Pitman cited an example for us of a woman who developed PTSD after being raped in an elevator. After a long period of avoiding elevators, even undertaking grueling 10-story walks up to her office, she eventually overcame her fear and started using elevators again.

"Fortunately, in this woman's case her prefrontal cortex was successful in saying to the amygdala, 'You're getting frightened now because you're thinking of getting into an elevator where this awful thing happened. But you also know all that is in the past; you're in the present now. And you're going to keep this under control and deal with it.'"

The principal goal when dealing with the recurrent episodes of PTSD is to prevent bodily arousal, according to Bessel van der Kolk, a psychiatrist at Boston University and clinical director of the Trauma Center at HRI Hospital in Brookline, Massachusetts.

"If the person with PTSD can keep his body from getting aroused, he can think more clearly and process the trauma. With his body calm and quiet, he can revisit the trauma with the therapist and move beyond it. If successful, the person gets the sense, 'This was something awful that happened to me but it's over now.' But if every time the person remembers the incident his body keeps pace with it and starts becoming freaked out again, then that person can never process the trauma and come to a sense of closure. In treatment we try get the person with PTSD to relive the experience while having different feelings about it."

Basically van der Kolk is teaching the PTSD victim to take advantage of his brain's plasticity and activate the power of imagination mediated by the frontal lobes. When the PTSD sufferer experiences himself as locked into a painful present, he loses perspective and, in van der Kolk's words, "behaves like a trapped animal." As an alternative he "teaches people with PTSD to read what their body is telling them. They learn to pick up the early warning signs of arousal and modulate them."

Zen Buddhist meditation is an example of body modulation, according to van der Kolk. The purpose of the meditation exercises is to gain perspective and distance on the trauma by remaining calm. "If a person remains calm enough, he or she can revisit and process the trauma, move through it. The big issue in treating PTSD is to keep people from going into the physiological arousal state where their thinking falls apart."

The "thinking" that van der Kolk is referring to emanates from activation of the prefrontal cortex. If the prefrontal cortex can be brought into play in the midst of a PTSD experience, the victim can imaginatively project himself into the future and envision an internal landscape no longer dominated by events from the past. "When the prefrontal lobes are brought into play, the person can simultaneously access past, present, and future images of himself. Thanks to the prefrontal cortex, it becomes possible to see beyond an unpleasant and painful present and transform it into positive terms. The whole picture is seen in context."

But not every sufferer from PTSD is so lucky as to bring the reaction under control with prefrontal effort. For the majority, the influence of the amygdala and the hormonal onslaught set off by the hypothalamus turn out to be too powerful. Medications are required to quell the turmoil of PTSD.

Block That Adrenaline   Neuroscientists have known for years that if a person is given a drug that blocks the action of adrenaline—one of the hormones released during a stress response—that person will no longer remember the emotional aspects of the upsetting experience. The same thing happens if the amygdala is damaged or destroyed. For example, one woman with brain damage restricted to the amygdala lost her ability to detect fear in other people's facial expressions. She also could not be fear conditioned. Neuroscientists believe that such fear extinction results from an absence of stimulation of the adrenaline receptors on the amygdala, where they are heavily concentrated. If so, a successful treatment might involve preventing the effects of adrenaline by blocking its receptors.

"In both laboratory animals and humans we have been able to block the formation of strong memories by giving drugs that block the action of adrenaline," McGaugh told us. "Termed 'beta-blockers,' these agents prevent the excessively arousing effect of emotional experience on memory. In the case of people with PTSD, the goal isn't elimination of the memory, but blocking its emotional over-expression, which lies at the basis of the disorder."

Treatment with beta-blockers aims at enhancing the capacity of the prefrontal cortex to overcome the amygdala-driven rush of emotions that herald the onset of a PTSD attack. After receiving the blocker, a person still remembers events but without the disorganizing emotional components.

Block Protein Synthesis   A newer and more controversial approach to treating PTSD is awaiting further development. It involves inhibiting the synthesis of proteins in the amygdala. Neuroscientists have long known that a certain amount of time is required before a memory encoding takes place within the brain.

"So after the rat has been conditioned to associate something unpleasant with a sound, we block protein synthesis in the amygdala. The next day the rat no longer has the memory. In just about every animal that's been studied you can block the formation of new memories if you block protein synthesis," according to Joseph LeDoux.

What's new in LeDoux's research findings is the concept of using protein inhibitors to block emotional responses *after* the memory has already formed. A precedent exists in humans for such an approach: Patients frequently forget a conversation that took place just prior to an electroconvulsive treatment (ECT). This happens because ECT interferes with the protein synthesis needed for memory consolidation of the conversation.

In this proposed new experimental approach to PTSD, the therapist reactivates the traumatic experience in the presence of protein inhibitors. The patient is encouraged to remember the event in vivid emotional detail. The treatment target is not the memory per se but the brain mechanisms responsible for forming the memory. The method works because memory retrieval and activation—"rehearsal" as James McGaugh puts it—makes the memory susceptible to disruption and manipulation. Such an approach has already been carried out in animals but awaits testing in people with PTSD.

As LeDoux was discussing eliminating symptoms of PTSD by interfering with protein synthesis, I recalled a recent cautionary letter to the editor of a medical journal about this research. In some instances, say, in Holocaust victims, the painful memories may over many years have become integrated into the person's identity. As painful and disturbing as the memories may be, additional risks may arise if the memories are erased. While the author of this cautionary suggestion has a point, memory eradication in some of these cases may still be necessary simply because some memories may be too painful to be endured. A good example is Primo Levi, author of two classic memoirs of his experience of the concentration camps. Despite valiant efforts to resist sinking into late-life despair in response to his repetitive and agonizing memories, Levi was eventually emotionally overwhelmed and committed suicide. Would Levi have been better off if his memories had undergone chemical ablation? With such a question we enter areas of inquiry and human concern that will forever remain beyond the scope of neuroscience.

What about Happiness?   Perhaps at this point, after reading about conditioned fear and PTSD, you're wondering,'What does all this have to do with the normal adult brain, the crown of creation? Can't neuroscience concern itself with anything happy? Is everything gloom and doom?'

Actually, neuroscientists are very interested in emotions like happiness, elation, and serenity. The problem is that the healthy, positive emotions don't command the attentions of those people who determine spending allocations for research dollars. Neuroscience is disease driven, with researchers concentrating on abnormalities and malfunctions. And since feeling good about yourself and the world around you isn't an impediment to your health and longevity

Images of Love

Being in love is one of the most exhilarating feelings in the human repertoire of emotions. In one study, fMRI scans were made of subjects who were deeply in love—first when they were shown photographs of their loved partner and again when shown photos of three friends of similar age, sex, and duration of friendship as the partner.

Although love and friendship are both positive emotions, the study showed distinct differences between them in the areas of the brain that were activated and deactivated, as seen opposite.

Strikingly, most of the deactivated regions coincide with regions that other studies have shown to be active in people suffering depression, sadness, or anxiety. This finding shows that activity in brain structures involved in negative emotions is suppressed in the state of love.

(in fact, such attitudes will increase your chances of living long and healthily, as we'll discuss in greater detail later), the healthy brain traditionally has gotten short shrift.

Nonetheless, we do know certain important facts about how the brain processes positive emotions like happiness. What's more, the tide is turning; neuroscientists now want to learn as much as possible about what makes people happy instead of concentrating exclusively on why they are made miserable by disease and ill health.

First of all, we know that no single area of the brain controls the expression of happiness, smiles, and laughter. And why should that be surprising? Happiness involves an internal state—a cozy, warm feeling—that is mediated by the limbic system, notably the amygdala and other structures connected to it. But happiness, at least in adults, also involves a cognitive component: comparing current conditions with previous ones; recognizing the fulfillment of personal desires; and taking appropriate steps to gain some measure of approval, support, and affection from other people. In order to carry out these cognitive operations, we have to employ our frontal lobes along with memory, mediating areas like the hippocampus, parts of the thalamus and, again, other parts of the frontal lobes.

Second, we know that happiness can't be reduced to a formula. Some people are happy even under circumstances that would test the fortitude of the vast

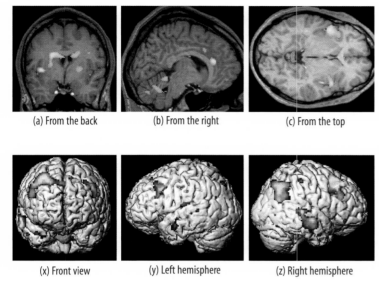

(a) From the back	(b) From the right	(c) From the top

(x) Front view	(y) Left hemisphere	(z) Right hemisphere

All regions activated by photos of the loved one are in the middle of the brain, as seen in three cross sections (top). These include (a) the caudate nucleus, putamen, and insula; (b) the anterior cingulate and cerebellum; and (c) the insula and hippocampus in both left and right hemispheres. The specific location and combination of these sites seems unique, suggesting that functional specialization in the cortex also applies to emotions.

Most of the regions deactivated when viewing photos of the loved one (bottom; x, y, and z) are on the surface of the brain, except perhaps the amygdala, a deep-brain nucleus. The deactivated regions also seem to be primarily in the right hemisphere.

majority of us (the late Mother Teresa comes to mind). In such instances, firmly held convictions, faith, or a belief in the worth of every human being can overcome external circumstances of poverty and deprivation that would drive a less spiritual person to despair. Will we ever turn up a part of the brain that will "explain" the happiness of a Mother Teresa? Personally, I doubt it. But perhaps we will be able to explain the neurological basis of optimism as opposed to pessimism. In fact, neuroscientists have already made significant advances toward that goal.

Richard Davidson is a professor of psychology and psychiatry at the University of Wisconsin, where he directs the brain-imaging laboratory. His research involves comparing people who always seem to be overly sad and pessimistic with the lucky ones who remain upbeat even under the most trying of circumstances.

"We have discovered that there are differences in certain circuits in the brain that differentiate between these positive, happy individuals compared to those who show more vulnerability in response to the emotional events in their lives."

Optimists versus Pessimists   Using PET scans and fMRI, Davidson has discovered increased activation within the left prefrontal cortex among individuals with a positive, happy outlook. They also show inhibition in the amygdala. Conversely, the amygdala is increasingly activated in unhappy, vulnerable people and there is an increase in activity of the right prefrontal cortex, not the left. As mentioned in the sidebar on page 32, Davidson's research is also proving helpful in assessing temperament in infants.

But temperament isn't a guarantor of happiness. One can be happy despite the recognition of loss. Indeed, such serenity is a form of happiness much prized by wise men and women throughout the ages. And although I have no doubt serenity will someday be correlated with specific chemical or electrical patterns in the brain, I still don't believe neuroscience will discover a neurological foundation for serenity or any expression of happiness. If it does, the next logical step will be pills or other technological aids toward furthering happiness. But that's unlikely to be helpful in the long run. We already know of drugs and chemicals that in the short run can free the mind from cares and usher in a state of short-lived happiness. But eventually the drug or the booze wears off and we wake up to the same problems that drove us to seek the fleeting solace of chemistry. No, happiness comes from a mix of attitude and behavior—lifestyle, as many of us would refer to it—that can't be reduced to a formula involving brain circuits or chemicals. Neuroscience is unlikely to be much help here, at least in the foreseeable future. Instead, it will remain most helpful in providing an explanation for happiness's dark mirror image: depression.

Debilitating Depression   While all of us experience days when we are "down," when nothing we do seems to turn out right and we feel discouraged and hopeless, a depressive illness is much more debilitating. Over the next century, depression will be the number one cause of disability in the developing world and the number four cause of death worldwide. Currently it afflicts 17 percent of people in the United States—12 to 13 percent of men and over twice as many women (about 25 percent). That breaks down into somewhere between 15 and 25 million Americans with a depressive episode in a given year.

Depression is particularly ominous because of its known association with suicide. More than 35,000 suicides per year in the United States establish it as the eighth leading cause of death. In fact, more people now die by their own hand than are killed by AIDS.

For more than a decade, Lauren Slater suffered virtually unrelieved depression. Since the advent of Prozac, she has been able to become a practicing psychologist and best-selling author.

A depressed person is also at increased risk for alcoholism, drug addiction, and a host of medical conditions such as heart disease and stroke. Nor is depression an equal opportunity affliction. Depression is more common among single, separated, and divorced people—particularly among elderly men, in whom the rate exceeds that found in adolescence.

Although certain components are common to the depressive experience (next page), they skirt a basic question: Is depression a disease? Most diseases are understood in terms of how and why they occur. But depression is thought to have many causes—some situational, some genetic, and some not attributable to anything in particular. For this reason, neuroscientists are more comfortable referring to depression as a syndrome—a constellation of symptoms and signs.

Writer and psychologist Lauren Slater compares depression to "one of those horrible tumors made up of nails and hair and all kinds of horrible parts. It's part terror and part deadness." Slater speaks from painful experience. A sufferer from depression since age 14, she has been hospitalized more than a dozen times and, in response to her suffering and despair, seriously considered suicide on several occasions. In her best-selling books, *Prozac Diary* and *Welcome to My Country,* Slater details depression's bleak and painful landscape:

"When you're depressed you're no longer able to control your own life. You're no longer in the driver's seat but your symptoms are driving you. There is a sense that you and depression have become one. To feel depressed, to me, is to feel

SIGNS OF DEPRESSION

Not everyone who is depressed experiences every symptom. Some people experience a few symptoms, some many. The severity of symptoms varies with individuals and also varies over time.

- Persistent sad, anxious, or "empty" mood
- Feelings of hopelessness, pessimism
- Feelings of guilt, worthlessness, helplessness
- Loss of interest or pleasure in hobbies and activities that were once enjoyed, including sex
- Decreased energy, fatigue, being "slowed down"
- Difficulty concentrating, remembering, making decisions
- Insomnia, early morning awakening, or oversleeping
- Appetite and/or weight loss or overeating and weight gain
- Thoughts of death or suicide; suicide attempts
- Restlessness, irritability
- Persistent physical symptoms that do not respond to treatment, such as headaches, digestive disorders, and chronic pain

dead. Time seems endless, horrifying; a minute seems like a week. It's the ultimate isolation and wordlessness—which is why it's so hard to talk about it. Thus depression is deadness, terror, horror, and distorted time."

Early Theories Explanations for depressions such as Lauren Slater's have varied over the centuries. Probably one of the most insightful comments came from a seventeenth-century sufferer, Robert Burton. In his *Anatomy of Melancholy*, Burton elaborated on a theory of causation traceable to the Father of Medicine, the Greek physician, Hippocrates.

Hippocrates postulated the existence of four humors: phlegm, yellow bile (also known as choler), blood, and the black bile of melancholy. According to the humoral theory, a person's temperament depended on the balance of the four humors within the body. Too much black bile resulted in the melancholic temperament, corresponding to what we would label today as depression.

Based on his own neglected childhood, Burton suggested that lack of affection in childhood could sometimes so warp a person's character that he or she could not feel or express love for self or others. The historian Jacques Barzun, writing in his epic *From Dawn to Decadence*, aptly captures Burton's concept of depression: "The melancholy individual is the plaything of opposite forces; he despises himself and then acts arrogantly; he is envious of others and knows that he is undeserving; he wants friends and lovers but does not know how to make the right approach and he alienates those who begin to feel affection for him."

Nobody today believes in "humors," yet contemporary beliefs about depression and its cure share similar assumptions. Instead of humors, psychopharmacologists (specialists in the treatment of mental illness by drugs) speak of neurotransmitter imbalances. Remember that neurotransmitters are the chemical units of commerce within the brain.

Although more than a hundred neurotransmitters have been identified, we know the most about only a handful of them: dopamine, norepinephrine, serotonin, and acetylcholine. The point of origin for these neurotransmitters is a small focus of cells, located in the brain stem, from which molecules ascend in a wide, fanlike pattern to influence large swaths of the brain. One or more of these circuits are involved in depression.

"We know that depression is a brain disease that involves those areas important in generating normal emotions," says Charles Nemeroff. "Included here are the cerebral cortex, the amygdala, the hippocampus, the hypothalamus, and other areas."

The initial suggestion that depression is a brain disease resulting from a chemical imbalance originated in India. In 1931, two researchers brought to world attention "a new drug for insanity." Reserpine, extracted from the root of the plant *Rauwolfia serpentina*, calmed psychotic patients but at a cost: drug-induced depression.

Eventually reserpine-treated animals were used as laboratory models of human depression. Certainly the animals mimicked many of the signs of human depression, such as psychomotor retardation, a 10-dollar term that essentially means the animals tended to stay put and displayed little interest or enthusiasm for anything going on around them.

Taking their cue from the animal findings, early psychopharmacologists set out to discover a drug capable of reversing reserpine-induced depression. If a drug could be developed that got the animals moving about again, the same drug just might bring about the same effect in depressed people—or so went the reasoning at the time. But such a drug eluded investigators. Finally, a clue emerged from an unexpected source.

Odd Side Effects   In 1952 reports began to circulate about an odd side effect associated with the new tuberculosis drug, isoniazid. Not only did their tuberculosis improve, but the patients given the drug also experienced a marked and sustained elevation in mood. Even more intriguing, a small number of the patients exhibited an extreme mood elevation (mania) that in some cases required psychiatric hospitalization. In short, more was going on than simple patient exuberance at having their tuberculosis brought under control. In search of an explanation, researchers turned once again to animal studies.

The neurotransmitters serotonin, norepinephrine, and dopamine are produced in several places deep in the brain. From their production sites they travel throughout the brain, as shown in simplified form here. Serotonin (red), which helps adjust body temperature and influences sleep and mood, and norepinephrine (blue), associated with the fight-or-flight response, go to the limbic system as well as throughout the cortex and into the cerebellum at the back of the brain. Dopamine (green) is also found in the limbic system and frontal cortex and in the basal ganglia, which help control skeletal muscles. It is mainly produced in an area called the substantia nigra. In Parkinson's disease, these cells are lost, resulting in a dopamine deficiency.

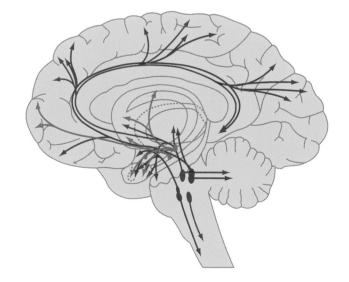

If a rat was given a dose of isoniazid prior to reserpine, the usual motor slowing failed to occur, suggesting that isoniazid exerted some chemical modification on the usual breakdown pathway of reserpine. Chemical analysis revealed that isoniazid and other drugs of the same family acted as inhibitors of an enzyme within the brain called monoamine oxidase (MAO). This enzyme breaks down the monoamine neurotransmitters dopamine, serotonin, and norepinephrine in the following manner:

After release from a neuron (the releasing neuron is called "presynaptic"), the monoamines wend their way across the synaptic cleft and attach to their specialized receptors on the postsynaptic membrane. After activating their receptor, the monoamines return to the presynaptic neuron (a process called "reuptake") for either repackaging to make new neurotransmitter molecules, or inactivation and breakdown into their constituent parts. But in the presence of an MAO inhibitor, the inactivation and breakdown don't take place. This prolongs the life of the monoamine neurotransmitters in the presynaptic neuron and eventually leads to an increase in the amount of the neurotransmitter in the synapse.

Neuroscientists now had their explanation for why reserpine causes depression and MAO inhibitors like isoniazid relieve the depression. Reserpine causes a lowering of the levels of the neurotransmitters by exposing them within the presynaptic neuron to the breakdown actions of MAO, resulting in a reduced amount for release into the synapse. MAO inhibitors neutralized this process and resulted in more neurotransmitter in the synapse.

Additional research revealed a second class of antidepressants that increase monoamine levels by a different mechanism: blockage of reuptake from the synapse into the presynaptic neuron. Ordinarily, after release from its receptor on the postsynaptic membrane, the transmitter is taken up by the presynaptic neuron to be either reprocessed or broken down by MAO. But in the presence of a reuptake blocker, the transmitter remains within the synapse for a prolonged period, thus lengthening and enhancing its actions. Drugs in this category were originally called "tricyclics," so named because structurally they consist of three fused rings.

A Two-Part Hypothesis   Taken together, these observations suggested a two-part hypothesis: Depression results from inadequate monoamine neurotransmission;

treatment works either by making more neurotransmitter available for release (the MAO class of antidepressants) or by blocking the reuptake of the transmitter from the synapse (the tricyclic class of antidepressants). The use of both classes results in an increase in the availability of the monoamine neurotransmitters within the synapse.

This monoamine deficiency theory of depression played a seminal role in the development of biological psychiatry. Initially, psychopharmacologists emphasized norepinephrine, and they synthesized new drugs aimed at raising norepinephrine levels in the synapse. More recently, the monoamine serotonin has captured the spotlight. Drugs such as Prozac, Zoloft, and Paxil inhibit serotonin's uptake, increase the level of that neurotransmitter in the synapse, and, in many cases but not all, counteract the symptoms of depression. For Lauren Slater, the serotonin reuptake inhibitor Prozac literally turned her life around. A chronically depressed person with a bleak future, Slater went on to become a successful writer and clinical psychologist.

Today the monoamine deficiency hypothesis is recognized as flawed. For one thing, although the chemical actions of antidepressant drugs occur soon after administration, improvement in the depression typically takes several weeks. This delay suggests that antidepressants bring about long-term modifications in brain function in addition to their immediate effect of increasing the amount of neurotransmitter in the synapse. The effect is similar to what happens when a new musician replaces a long-established musician in a string quartet: it takes time for the other musicians to adjust their playing to the "style" of the newcomer. Similarly, the different neurotransmitters and other chemicals co-exist in a delicate balance within the brain; altering one of them brings about, over time, changes in the others.

Multiple Targets   Some of the most recently developed antidepressants target several neurotransmitters rather than only one. This approach is tacit recognition that no one is certain about the neurochemical details of depression. In addition, the culprit responsible for depression may vary from one person to another. This would be consistent with observable differences in depression. Not all depressions look the same: crying and expressions of hopelessness in some; lack of energy, sleeplessness, and loss of appetite in others. To bring order into a rather disordered field like psychopharmacology, neuroscientists must concentrate on what's happening in the brain of the depressed person.

Helen Mayberg, a psychiatrist and neurologist at the University of Toronto, carried out PET scans of depressed people to take a closer look at the brain in depression. A key observation underlay Mayberg's investigations. Depression isn't marked simply by feelings of sadness, pessimism, and despair but also by specific problems with thinking. For instance, the depressed person may be apathetic, disinterested, and disconnected.

"In depressed people the turnoff switch by which thinking controls emotions isn't working properly. As a result emotion overrides thinking," Mayberg says. Remember that the frontal lobes are responsible for thinking, aspects of memory other than initial encoding, and higher-order mental processing in general.

"We encounter in depression the same kinds of problems that result from frontal lobe damage. Depression results from a person's inability to maintain a balance between the thinking and the emotional centers," according to Mayberg. "The balance is tipped in favor of the emotional centers. It's as if the thinking parts of the brain are shut down."

To test this hypothesis, Mayberg compared PET scans of depressed people at three key points: prior to any treatment, after three weeks of antidepressant treatment with Prozac, and a final scan three weeks after that.

"In a person suffering a major depression we initially see underactivity in the frontal lobes that correlates in severity with that of the depression. Those parts of the brain that process emotion, in contrast, are working in overdrive. With a PET scan we can actually visualize this shift in the normal balance: an increase in activity in the emotional areas, and a dampening of activity in the thinking centers."

Changing the Balance in the Brain   After one week on an antidepressant like Prozac, the PET scan changes. While not a lot is happening in the cortex, activity in the limbic-emotional areas is, in Mayberg's words, "being recalibrated, retuned" toward less activity. After six weeks, the frontal cortex on both sides shows increased activity, almost to normal level.

"The treatment has changed the balance in the brain. The antidepressant has turned down the emotional centers and turned up the cortical centers. We think that this rebalancing results primarily from the effect of the antidepressant on those deeper-lying emotional centers."

Mayberg's PET scan studies are consistent with the observations of neuropsychiatrists that patient improvement lags behind chemical brain changes. Although the administration of Prozac leads to a fairly immediate increase in brain serotonin, several weeks usually go by before patient and relatives begin noticing changes in mood, thinking, appetite and sleep. To explain this delay, neuroscientists focus on the phenomenon of the "cascade."

Enhancing serotonin is just the beginning. The rise in serotonin levels sets off a cascade of chemical events that change the way brain cells talk to each other and ultimately result in a change in the balance of the frontal and emotional centers. The cascade of chemical reactions has altered molecular processes in the brain.

Changes in molecular processes and changes in "talk" between brain cells no doubt account, as well, for the high recurrence rate of depression. Each recurrence may be the result of previous damage to the brain, with additional damage resulting from each episode of depression. What's the evidence for this?

As noted in the discussion of PTSD, stress-related problems in memory result from damage to the hippocampus, where memories are initially encoded. And since depression is also stressful, it should come as no surprise that the MRIs of depressed people show decreases in hippocampal size. This shrinking of the hippocampus results from the high levels of stress hormones that, for reasons that aren't completely understood, wreak their greatest damage on that brain site.

"We don't know at this point," Mayberg says, "if depression by itself damages the hippocampus or if the damage only occurs in depressed people with poorly regulated stress responses. In either event, the resulting hippocampal damage impedes that structure's ability to regulate additional stress. This results in a vicious cycle. Depression elevates stress; stress hormones damage the hippocampus; the damaged hippocampus in turn damages those brain areas needed to modulate future stress."

Antidepressant drugs provide one approach to the vicious cycle. Raising the level of serotonin or other neurotransmitters helps to contain the stress response

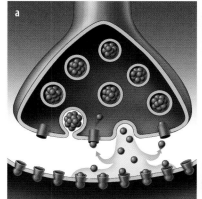 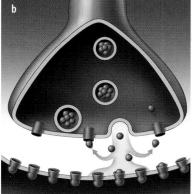 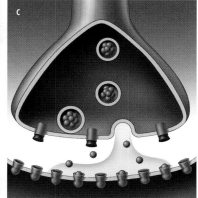

Depression at the Cellular Level In the normal brain (a), serotonin is stored in tiny sacs called secretory vesicles at the end of nerve cells. An electrical signal triggers the release of serotonin molecules, which travel across the synapse—the minute gaps between nerve cells—and bind to receptors located on the surface of the other nerve cell. After signal transmission is completed, the receptors release the serotonin back into the synapse, where it is either reabsorbed by the first nerve cell for later use, or broken down and disposed of by enzymes.

In the brain of the depressed person (b), serotonin levels are low, despite a normal number of receptors. Some studies also suggest a link between low serotonin levels and impulsive behavior, eating disorders, sleep disorders, and difficulties processing emotions.

In (c), the antidepressant medications known as serotonin-reuptake inhibitors work by blocking the reabsorption of serotonin by the signal-sending neuron, thus increasing the amount of serotonin available to the brain.

and decreases the resulting damage induced by those elevated stress hormones. Whichever neurotransmitter is targeted, the effect is the same: a chemical cascade that changes the chemical balance of many other neurotransmitters.

"We know," explains Mayberg, "that if we increase the amount of available serotonin we change the chemical balance within the brain and convert a poorly functioning brain into a normally functioning one. Over several weeks metabolism increases in the frontal areas and decreases in the emotional centers."

Talk Therapy But medications aren't the only approach to treating depression. Another approach replaces chemicals with the most time-honored treatment of all: talk. In cognitive therapy, the depressed person is encouraged to use the thinking aspects of his brain to wrest power back from the emotional centers. For instance, the depressed person is asked to reexamine deeply held assumptions about himself. Typically, a depressed person thinks of the world as a threatening place; thinks of himself as unable to bring about positive changes. But even under circumstances when change is difficult or impossible, the depressed person still has the potential to control one thing: his personal responses.

The Cycle of Bipolar Disorder

Also called manic-depressive illness, bipolar disorder is a type of mood disorder characterized by severe highs (mania) and lows (depression). Sometimes the mood switches are dramatic and rapid, as seen in the scans at right of someone who went from depressed to manic and back in 10 days. More often the change is gradual. Someone in the depressed phase can have any or all of the symptoms of depression (page 136). In the manic phase, he or she may be overtalkative and have a great deal of energy. Composer Robert Schumann, for instance, was prodigiously productive during his manic phases. However, mania often affects thinking, judgment, and social behavior in ways ranging from unwise business decisions or romantic sprees to actual psychosis in some cases.

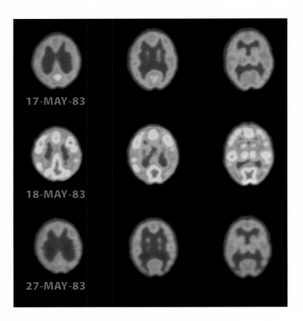

17-MAY-83

18-MAY-83

27-MAY-83

In cognitive therapy, the depressed patient works at strengthening the frontal lobe pathways—the thinking pathways—by learning to recognize those thoughts that precede or accompany depressive feelings, and alter them. For instance, the depression-inducing thought "I *always* screw things up" would be replaced with the more reasonable thought "I *sometimes* screw things up." By a gradual process and over many months, more accurate, less self-abasing points of view replace self-defeating inner dialogues involving futility and hopelessness. Or attention is shifted from negative thoughts that induce negative feelings to positive thoughts that foster positive feelings.

What makes cognitive therapy work? At this point neuroscientists aren't certain. Helen Mayberg explains this therapeutic technique in neurological terms. "While drugs work on the emotional areas deep in the brain, cognitive therapy exercises the thinking areas of the brain and thereby affects the balance from top down. Cognitive therapy exercises the cortex and thereby strengthens the pathways by which the thinking brain influences the emotional brain. If depression results from a negative balance between the thinking and feeling parts of the brain, cognitive therapy restores a normal balance."

Although neuroscientists have made great progress over the past five decades in understanding depression, many unanswered questions and issues remain. For one thing, researchers have dedicated most of their efforts to only a handful of

The effects of composer Robert Schumann's bipolar illness are strikingly evident in a chart of his productivity over some 30 years. In two years when he suffered manic phases he created two dozen or more musical works, as compared with no output during a year of severe depression.

Schumann's musical works (catalog numbers) charted by year of composition:

- **1829:** 007
- **1830:** 001
- **1831:** 008
- **1832:** 002, 003, 004, 124
- **1833:** 005, 010
- **1834:** (no output)
- **1835:** 099
- **1836:** 009, 011, 022
- **1837:** 013, 014, 017
- **1838:** 006, 012
- **1839:** 015, 016, 018, 021, 032
- **1840:** 019, 020, 023, 025, 026, 027, 028, 029, 030, 031, 033, 034, 035, 036, 039, 040, 042, 043, 045, 048, 049, 051, 053, 057, 077, 127, 142
- **1841:** 024, 052, 054, 064, 120
- **1842:** 037, 038, 047
- **1843:** 041, 044, 050
- **1844:** (no output)
- **1845:** 046, 056, 058, 060, 072
- **1846:** 055, 061
- **1847:** 059, 063, 065, 080, 084
- **1848:** 062, 068, 071, 081, 115
- **1849:** 066, 067, 069, 070, 073, 074, 075, 076, 078, 079, 082, 085, 086, 091, 093, 094, 095, 098, 101, 102, 106, 108, 137, 138, 141, 145, 146
- **1850:** 083, 087, 088, 089, 090, 096, 097, 125, 129, 130, 144
- **1851:** 100, 103, 104, 105, 107, 109, 110, 111, 112, 113, 117, 119, 121, 128, 136
- **1852:** 122, 135, 139, 140, 147, 148
- **1853:** 114, 118, 123, 126, 131, 132, 133, 134, 143
- **1854–1856:** (no output)

Timeline: 1829 1830 1831 1832 **1833** 1834 1835 1836 1837 1838 1839 **1840** 1841 1842 1843 **1844** 1845 1846 1847 1848 **1849** 1850 1851 1852 1853 **1854** 1855 **1856**

- ↑ 1833 — Suicide attempt
- ↑ 1840 — Hypomanic throughout 1840
- ↑ 1844 — Severe depression throughout 1844
- ↑ 1849 — Hypomanic throughout 1849
- ↑ 1854 — Suicide attempt
- ↑ 1856 — Died in asylum (self-starvation)

the hundred or more neurotransmitters within the brain. Surely these other neurotransmitters must be playing some role in depression. Besides, altering one neurotransmitter leads to alterations in many others. Thus, it's unlikely that a single neurotransmitter or receptor causes depression—or any other brain disorder discussed in this book. Elliot Valenstein makes this point in his book *Blaming the Brain*:

"The belief that the complex cognitive and emotional states that underlie any mental disorder are regulated by a single neurotransmitter … is probably no more valid than the idea held earlier by phrenologists who believed that complex mental attributes could be localized in one specific part of the brain."

An Emotional Deficiency   So far, we have focused on disorders that involve excesses of feeling and emotion. Yet other disorders involve just the opposite: a *deficiency* of emotions. Perhaps you know a person with such a disorder. If so, he or she undoubtedly strikes you as remote, detached, cold, perhaps even cruel. Typically, we refer to such a person as lacking in empathy. As an example, meet Marvin Bateman.

Twenty-three years ago Bateman, now 57, suffered a stroke that destroyed parts of his right hemisphere (specifically, the right somatosensory cortex). As a result of his injury, Bateman experiences great difficulty empathizing with others.

The Story of Phineas Gage

In 1848, a young foreman on a railroad blasting crew suffered a brutal accident: A 13-pound tamping rod shot through his left cheek, behind his eye, and up through his left frontal lobe. Exiting his skull, the rod flew another 50 feet through the air. Miraculously, Phineas Gage survived, and with few physical problems except for having lost the sight in his left eye. But his personality was forever changed. Once polite and proper, he became by turns obstinate, childish, and crude, prone to uttering gross profanities.

One of the two doctors who had been called to the scene of the accident kept track of Gage's behavior and concluded that the accident had destroyed "the equilibrium, or balance, between his intellectual faculties and animal propensities."

Unable to get his old job back (employers and friends shunned him), he drifted into such jobs as a stable hand and coach driver. He also appeared with P. T. Barnum's circus as a sideshow freak. In 1861, nearly 13 years after the accident, Phineas Gage died in San Francisco at the age of 37.

Once a loving husband and doting father, he is now unable to cry at a funeral or enjoy the successes of his children. As a result of his emotional disconnection from other people, Marvin can no longer hold a job or even make everyday decisions by himself.

"Mr. Bateman suffered damage to the parts of the brain that allow him to have a full set of feelings, to internally re-create and simulate the feelings of others," according to Antonio Damasio, Bateman's neurologist at the University of Iowa. "Since he cannot automatically generate feelings that match the feelings of others, his capacity for empathy is lost."

A Crucial Distinction   To explain what's happening with Marvin Bateman, Damasio makes a crucial distinction between emotions and feelings. For most of us, emotions and feelings are the same, and any distinctions between them seem trivial. But they are not the same, according to Damasio, and if we don't take the trouble required in making the necessary distinctions, we won't understand what has gone terribly wrong in the mental life of Marvin Bateman.

"Emotions are public and can be seen. While you can see me crying or getting angry, for instance, you can never see my feelings. You can never see what I'm experiencing when I feel angry, happy, or sad. Those feelings are private, internal, and personal."

As an example of Damasio's point, think for a moment of a situation in your recent past when you got angry. Perhaps it was yesterday morning when your neighbor extended a lukewarm curt greeting, or maybe no greeting at all. Make a mental note of your feelings now as you think back to such a slight. Next,

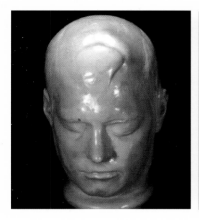

Phineas Gage's death mask (far left) seems remarkably tranquil considering the hole in his skull (near left) made by the tamping rod that destroyed his frontal lobe

bring to mind something that made you happy. You might want to remember something like a relaxing afternoon spent at the beach. Note that feeling as you experience it and contrast it with the previous one. How would you describe the differences in these two feelings?

If you probe deeply enough, you'll discover that the differences in your feelings involve different bodily states. One recollection (the rudeness of your neighbor) arouses an uncomfortable, constricting feeling, while the beach memory arouses a feeling of expansive relaxation. Those two very different experiences provided the raw material for your emotions that, at the time, were visible to the acute observer. Anyone observing your facial expressions during the two very different experiences would be in a position at that moment to "read" your emotions and know that you were emotionally upset after encountering your neighbor and contented while lying on the beach. But the feelings that accompany your present *recollection* of those two previous events aren't available to anyone observing you. Only *you* have access to these feelings and you experience them in terms of internal bodily states.

"Think of the body as the theater of the emotions," suggests Damasio. "Something happens that triggers the bodily responses, and these responses (blushing, a raised angry voice, etc.) characterize the different emotions. But then a second, very important process must take place. The emotion must be mentally assimilated and represented. This involves an entirely different set of brain structures that help maintain the impact of the original emotion. Why is that important? Because you will be able to use your feelings for future planning, anticipation, and decision making. It will also enable you to empathize with people experiencing the same feelings."

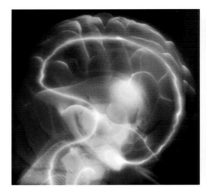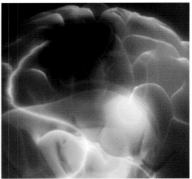

For Marvin Bateman (opposite), the normal neuronal communication circuit among the various parts of the brain involved in emotions was broken 23 years ago when a stroke destroyed the somatosensory cortex in his right hemisphere.

Brain damage such as that incurred by Marvin Bateman results in a split: He can still experience an emotion (if you slap him on the face he will get appropriately angry), but that experience will not be encoded in his brain in the form of feelings that will be available to him to influence his future behavior.

"While he can experience an emotion in the present, he cannot later connect a similar situation with the appropriate feeling. As a result, the emotional system and the fact system remain separated. Mr. Bateman can't mesh them together in his own brain. As another consequence of his brain damage, he's unable to appreciate the emotions of others. For instance, if another person expresses a certain emotion that, normally, is accompanied by a certain feeling, it will be more difficult for Mr. Bateman to get into that person's mind and have a sense of what that person is feeling. 'Empathy' is more difficult in a person with the kind of brain damage suffered by Mr. Bateman."

During our visit we witnessed an example of what Damasio was talking about. Marvin and his wife, Arlene, were looking at pictures taken before and after Marvin Bateman's stroke. Struck by the contrast between Marvin then compared to Marvin now, Arlene began to cry softly. Marvin, in contrast, remained curiously unmoved, as if unable to fully appreciate the full significance of her sorrow. He appreciated the fact of the situation—recognized that his wife was crying—but his awareness stopped there.

Arlene recalls the first occasion when she became aware of her husband's emotional incapacity: "After I brought him home from the hospital I said to him 'If I had been the one who suffered the stroke, would you take care of me?' He thought for a moment and then said, 'I probably shouldn't say this, but I don't think I would.' I knew then that the emotional connection between the two of us was no longer there. If I try to talk to him about something that's bothering me, he's just not there for me. He's just not there for anybody when they get upset. He has a

As a result of his injury, Bateman cannot empathize with others. "He's just not there for anybody when they get upset," says his wife, Arlene. "He knows that he should feel something, but he doesn't. It's as if a cord has been cut."

hard time showing any type of sympathy. 'I wish I could cry,' he once told me. But he just can't. While in his mind he knows that something sad is happening, he just can't get to the feelings. At a funeral, for instance, he didn't show any emotion at all but spent all his time wandering about and talking to various people. He knows that he should feel something, but he doesn't. It's as if a cord has been cut."

Antonio Damasio explains: "Arlene's crying doesn't create a repercussion in him at the level of *feeling*. He recognizes that she is crying but that recognition is not accompanied—as it would be in a normal person—by a feeling of sympathy and empathy for the person who is crying."

Lacking that sympathy and empathy, Bateman has great difficulty knowing how to respond to other people's emotional expressions. While most of us would make some attempt to comfort a grieving person, Bateman under the same circumstances seems confused about how best to proceed. He doesn't intend to be cold or uncaring. It's just that as a result of his brain injury he can't arouse within himself the feelings he had prior to his stroke on those occasions when he experienced grief. Lacking this capacity to access the *feeling* of grief within himself, he can no longer empathically recognize it in others.

Such impairment also exerts a devastating effect on the person's ability to work and get along with others. "To hold any kind of job—to maintain any kind of human relationship—you're constantly dealing with aspects of emotion such as reading the emotions of others from their facial expression or their tone of voice," Damasio told us.

In Marvin Bateman's job as a heating contractor he experienced problems "reading" the emotional state of his customers and adopting the appropriate demeanor and response. As a result of these failures of emotional resonance —failures to *feel* the situation from the other person's point of view—Marvin Bateman discovered that he could no longer hold a job.

Although Marvin Bateman's injury involved his right parietal lobe, a similar picture can result from damage in the prefrontal cortex, which links emotions with events and decisions. "It really doesn't make any difference whether the damage is in one place or another. In either case the person can no longer make the proper link between situations and the feelings that ordinarily accompany them," according to Damasio.

Indeed, when dealing with emotions and feelings it's best to think in terms of systems rather than discrete components. Moreover, different components of the system may be involved in mediating specific emotions. For instance, neurobiologist Andrew Calder describes a patient (whom he calls "NK") with loss of his ability to register disgust. The patient cannot recognize disgust in others from facial expressions, nonverbal emotional sounds ("ugh," "yuck"), or tones of voice ("You're going to eat THAT!"). PET scan images of NK's brain showed damage to two specific areas (the insula and putamen). It's likely that other brain areas mediate other emotions, such as happiness and sadness. Remember, depression is characterized by involvement of the frontal lobes, especially the right. Says Damasio, "We have something like a dozen different sites that form circuitries. The prefrontal cortex triggers the amygdala that triggers the brain stem or sends signals to change the somatosensory cortex, as with Marvin Bateman, or the cingulate and so on. It's a very loopy system."

Integrating emotions with thoughts can be difficult even in healthy adults. But thanks to the insights provided by neuroscience, we now know that integration *must* be achieved if we are to reach our full potential.

Joseph LeDoux: "As things now stand, the amygdala has a greater influence on the cortex than the cortex has on the amygdala, allowing emotional arousal to dominate and control thinking. At the same time, the cortical connections with the amygdala are far greater in primates than in other animals. This suggests the possibility of a balance in these nerve pathways. With increased connectivity between cortex and amygdala, cognition and emotion might begin to work together rather than separately. Thus, the struggle between thought and emotion may ultimately be resolved, not by the dominance of emotional centers by cortical cognitions but by a more harmonious integration of reason and passion."

As the Brain Enters Old Age

The furrows and folds of the brain in old age represent our unique history—a reflection of all that we are and all that we have been. Indeed, all that we know and can remember is etched in those hundred billion or so neurons that span the decades from birth until death. Whereas other cells throughout the body die off and are replaced, some of the brain cells present at our birth endure over our entire lives. This endurance becomes even more remarkable in light of the fact that during its final decades the brain may be tested by a new set of challenges: the slow but steady erosion of its complex structure by injuries or crippling degenerative diseases. But all is not gloom and doom. As neuroscientists unravel the secrets of the aging brain, they are learning that there is good reason for confidence and optimism. An organ long considered defenseless before the onslaughts of time, the brain is now recognized as capable of marshaling surprising powers of renewal.

Second Flowering

THE AGING BRAIN

Memory becomes especially important in old age. An intact memory enables the older person to examine the totality of his or her life and conduct that exploration that T. S. Eliot characterized as "to arrive where we started and know the place for the first time."

Indeed, memory is the most personally defining of all of the operations carried out by the brain. When we recall an event from our distant past we live it again. We can do this because of our brain's capacity for storage and retrieval in the form of distinct recollections. Memory virtuosos like the early twentieth-century French novelist Marcel Proust re-create events in such exquisite detail that, in some cases, the past exerts a more powerful and compelling influence than anything happening in the present. But even for the rest of us, who are not memory virtuosos, memory is no less important. We rely on our memory in ways that we don't even suspect unless we come into contact with someone who has lost his memory.

If memory fails we become a prisoner of the present, incapable of linking our current life with events from our past. And if we suffer from dementia (formerly called senility), our memory loss may be severe enough to transform our spouses, relatives, and friends into strangers. Truly, to arrive at this sad state is to lose what the philosopher Arthur Schopenhauer considered the singular benefit of old age: to see "life whole and know its natural course."

But memory is not the only function that may change with age. An older person's thinking may also begin to lack vigor and clarity. He or she may lose interest in events and people—in essence, that person beats a slow and measured retreat from the world. Such regressions don't occur to every older person, however. Why does one person retreat into darkness while another remains vital and wise?

Take America's former poet laureate, Stanley Kunitz, for example. At 95, Kunitz continues to write poetry when he's not working in his garden at his summer home in Provincetown, Massachusetts. During the filming of the documentary we attended one of Kunitz's poetry readings, where we encountered firsthand the undiminished power of the older brain:

> Light splashed this morning
> on the shell-pink anemones
> swaying on their tall stems;
> down blue-spiked veronica
> light flowed in rivulets
> over the humps of honeybees;
> this morning I saw light kiss
> the silk of the roses
> in their second flowering,
> my late bloomers
> flushed with their brandy.
> A curious gladness shook me.
>
> . . .
>
> I can scarcely wait until tomorrow
> when new life begins for me,
> as it does each day
> as it does each day.

A prime example of someone whose brain remains intact and vigorous in old age is former poet laureate of the United States Stanley Kunitz, shown above in 1962, at age 57, when he was the first poet in residence at the University of Arizona's Poetry Center, and at left in his garden in Provincetown, Massachusetts. In 1959, Kunitz was awarded the Pulitzer Prize for his book *Selected Poems* and has enjoyed a long and varied career as a poet, editor, essayist, and translator.

Later, we asked Denise Park, a research scientist at the Center for Aging and Cognition at the University of Michigan in Ann Arbor, about "graceful agers" like Stanley Kunitz. Although genetics undoubtedly plays a role, Park believes that "keeping the brain active and performing whatever tasks are needed to maintain intellectual vitality can increase the chances for optimum functioning." And since not everyone is a poet, these tasks will vary from person to person depending on occupation and life circumstances. It's the brain's plasticity that gives each person the opportunity to become a "graceful ager."

Increasing Variability   One manifestation of that plasticity is the great variability from one older person to another. According to Marilyn Albert, of Harvard Medical School, "This variability is one of the hallmarks of aging. If you look at a group of healthy 30-year-olds, they're very similar to each other in terms of brain functioning. But among older people, the differences between one person and another become greater."

Variability within the brains of older people continues into the ninth decade or later. "Aging is not a sudden event. It's a continuous event, a continuous process that in the normally aging person begins in the 20's and continues into late adulthood at a relatively steady pace," Denise Park says.

The Workings of Memory

Whatever we remember—whether it's how to tie our shoelaces or drive a car, the words to the "Star-Spangled Banner" or the birthdays of all our grandchildren—that memory is the product of molecular processes that strengthened synaptic connections involving several regions of the brain. Some memory traces are fleeting—the phone number of a restaurant in a city we're passing through that we remember just long enough to dial. Other memories, those with a strong emotional association, whether positive or negative, can last a lifetime.

Neuroscientists have found that different types of long-term memory are supported by distinct regions and structures in the brain. "Knowing how" is often called procedural or implicit memory. Riding a bike, playing tennis, walking, and talking are skills and habits we learned through repetition and now perform without conscious thought. Largely involving the cerebellum and motor cortex, they are also difficult to explain verbally. Automatic emotional responses—the panic attack of someone with PTSD, for example—are another kind of implicit memory, formed with strong involvement of the amygdala.

"Knowing that"—also called declarative or explicit memory—includes the facts and events that make up our lives. Critical to the formation of explicit memory are the hippocampus and the medial temporal lobe. The frontal lobe is also important for developing strategies to transfer information to long-term memory from the short-term and working-memory stores.

Short- versus Long-Term Memory Information enters the cerebral cortex through our senses and is held for a fraction of a second unless we pay attention long enough for it to transfer to short-term memory. Short-term memory is temporary storage with limited capacity (about seven items). Working memory is the strategic manipulation of that information.

To consolidate the information for long-term storage requires repetition or the application of meaning, both of which seem to trigger the protein synthesis that locks in what would otherwise be only temporary neuronal connections.

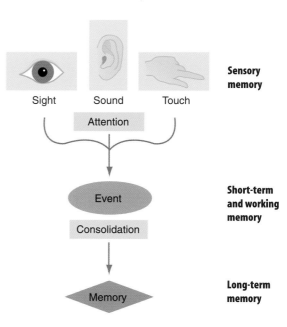

Sight   Sound   Touch

Sensory memory

Attention

Event

Short-term and working memory

Consolidation

Memory

Long-term memory

Most of what we commonly associate with "memory"—all the names, facts, and figures we have learned through life—involves structures on the inner, or medial, side of the temporal lobe, including the hippocampus and amygdala, as well as the specific sensory pathways by which the information enters. Nondeclarative, or procedural, memory (skills and habits) involves specific sensory and motor pathways as well as the cerebellum, the amygdala, and other structures deep in the white matter of the cortex. In both types of memory, the thalamus provides the gateway for sensory information and focused attention.

Working memory, an important window into our consciousness, occurs in the prefrontal cortex, just behind our high forehead. Short-term memory stores reside nearby in the frontal cortex. Working memory is what we use to remember a decision long enough to carry it out and where we hold different kinds of information simultaneously—the way an air traffic controller, for example, keeps track of arriving and departing planes as well as those that are in holding patterns, while simultaneously making and communicating decisions about each plane.

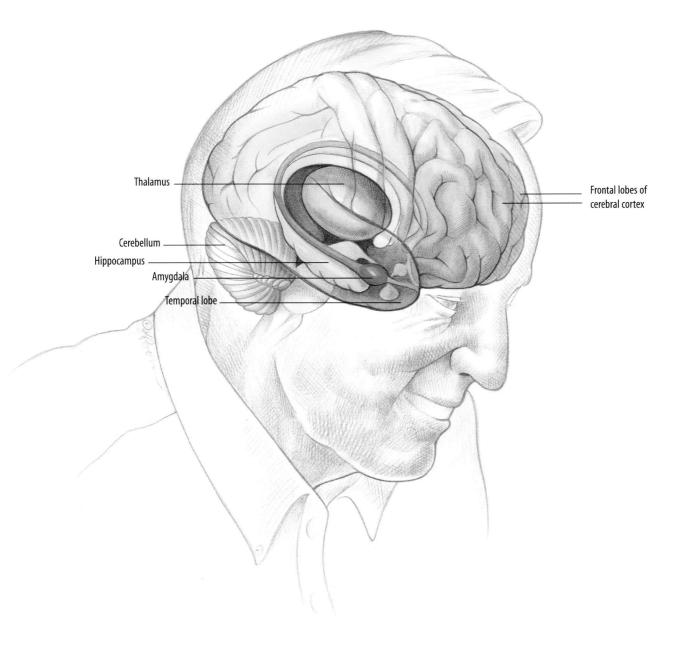

Thalamus

Cerebellum

Hippocampus

Amygdala

Temporal lobe

Frontal lobes of cerebral cortex

Aging begins in the 20's? Try telling that to the next 27-year-old entrepeneur or graduate student you encounter. Yet, it's true. Park's research shows that even when a person ages from 20 to 30, he or she is a little bit slower. Information is processed more slowly; held for a shorter time in conscious awareness; recalled less efficiently. These changes are incremental and begin as early as the 30's.

The Crowded Mental Desktop   As we grow older, changes in brain performance initially occur in our memory. Starting at late middle age, more time is needed to learn new information. Coupled with this slowdown, working-memory capacity—the amount of information that can be comfortably fitted onto the "mental desktop" for immediate retrieval and use—becomes more limited. Think of an overcrowded desk with all the items in comparative disarray. As a result of the crowding and clutter, long-term memory becomes less reliable, more time is required to enter new information into long-term storage, and more time is required to retrieve that information.

And we become more distractible as we age. For instance, a person in his or her late 60's or early 70's often finds it slightly harder to focus, keep to a conversational "point," and filter out extraneous noise. To get a feeling for the practical implications of these limitations, consider the experiences of a 75-year-old grandfather, let's call him Thomas, as he sits in a noisy restaurant with his grandson, Brent.

First, Thomas perceives that the waitress speaks "too fast" while announcing the dinner specials and he can't react quickly enough to catch all the items. Rather than ask the waitress to repeat the list, he decides to order from the menu. But it takes him slightly longer to read through the menu, and his grandson's scowl of impatience makes Thomas feel rushed and flustered, further prolonging the moment when he announces his selection.

With the orders placed, Thomas and Brent begin a conversation about some investments. This topic, suggested by Brent, makes Thomas slightly uncomfortable. That's because over the past several years he's experienced difficulty mentally juggling all the factors required for an intelligent discussion about investments. This difficulty would disappear if their conversation were taking place in the quiet of Thomas's home where, in the absence of distractions, he could spread

Dendritic length
(microns)

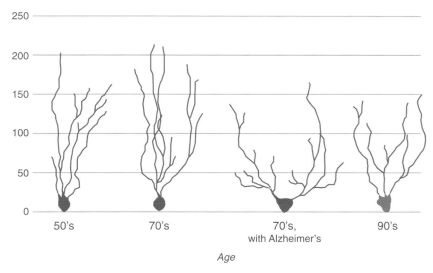

Normal Aging versus Alzheimer's Between the ages of 50 and 70, as the brain loses neurons in the normal course of aging, the remaining cells send out longer dendrites, compensating some-what for the loss of connections. The neurons of a 70-year-old Alzheimer's victim, however, typi-cally show no such growth. Indeed, they are equivalent to those of a healthy 90-year-old, whose neurons have begun to shrink in any case.

the relevant papers on the kitchen table. In addition, Thomas is experiencing great difficulty concentrating on Brent's comments against the blare of back-ground music coupled with the distracting effect of several conversations from nearby tables.

Thomas's difficulties illustrate the liabilities that, to a greater or lesser extent, can affect the older brain: slower processing of information, loss of the ability to consider more than a few items simultaneously, dulling of long-term memory, oversensitivity to background noise. All these factors contribute to the mistaken impression that Thomas is no longer capable of engaging in a coherent, focused conversation. Of these mental changes accompanying aging, distraction is the most immediately disruptive.

As Park reminded us during our conversation, "Aging is not like falling off a cliff." We don't perform perfectly normally until our 60's or 70's and then suddenly experience a dramatic falloff in our mental abilities. Rather, the decrements are subtle and creep up on us over many decades. Park attributes much of the blame for this creeping memory failure to information overload.

"The information-processing demands that occur in middle age are quite profound. Middle-aged people are always coming up to me and saying, 'I can barely remember what I did yesterday.' But it turns out they did hundreds of things yesterday. That's why older people actually are often better at remembering what they did yesterday than middle-aged people, who forget because on most days they're doing too much. They complain that they have a bad memory, yet the fault isn't with their memory but with their overcrowded schedules. When considerations like overload are factored out, however, the older person still experiences greater difficulty remembering names and everyday details."

Frontal Cortex Changes   In the case of Thomas's difficulties in the restaurant, most of them can be traced to changes in his frontal cortex. With aging, the frontal cortex is less able to sustain a sufficient working memory: the ability when confronted with delay or distraction to keep several items of information online at the same time. In some situations, working memory is more than just convenience and is absolutely essential for success, sometimes even survival.

For instance, imagine that you've flown to a new city and decide to rent a car. Within a short period of time you have to figure out the location of the rental desks, select the size and model of your car, arrange the appropriate rental terms (will you purchase insurance coverage?), quickly learn how to operate the vehicle, and rapidly scan a map of the route to your destination. En route, you have to decide about directions, choices at major intersections, and the speed you must maintain to keep to your estimated time of arrival.

"That's a great example of a daunting working-memory test," according to Park. "It involves rapid decisions, speed of processing, and a lot of active manipulation of information. Many 70-year-olds would avoid such a situation because it places too much stress on the working-memory capacity of the frontal lobes."

Reduced to a single word, the frontal lobe is concerned with the *manipulation* of information. Manipulation is not the same as registration of information. For instance, older people do as well as their younger counterparts on tests involving nothing more than information registration. In one of Park's tests measuring registration, the subject looks at a picture for about six seconds. He or she then immediately looks at a small picture fragment and decides by a simple Yes or No if the fragment was a component of the original picture. Age has no effect on performance in this test of passive memory, which doesn't involve any manipulation of information.

In a variation of the picture test that older subjects find more difficult, the picture is seen for only two seconds, followed by a four-second delay. During this delay the subjects are asked to maintain a "mental image" of the picture. After the four seconds they see a picture fragment and respond as to whether it was part of the original picture.

"We're interested in how the activation patterns of the older brain differ from what we observe with the younger brain," Denise Park explains. "And the fMRI patterns are quite different. When looking at the picture, young adults primarily activate their right visual cortex and right hemisphere, in general. Older adults, in contrast, activate both sides of their brain about equally." Other differences between young and old are also clear in scans such as those shown below.

On other tests for working memory, such as manipulating a list of words, young adults usually engage only the left frontal cortex, while older adults may

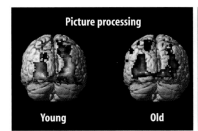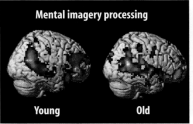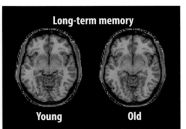

| Picture processing | Mental imagery processing | Long-term memory |
| Young Old | Young Old | Young Old |

Changing Patterns In general, older adults seem to process information differently than younger people. When looking at pictures, a task that involves the back of the brain, where the visual cortex resides (above, left), the pattern of activation in young adults is more focused on the right side, the side of the brain that processes visuospatial information. By contrast, the pattern in older adults is roughly the same in both the left and right sides of their visual cortex.

When asked to keep a picture in their mind for four seconds (middle, view of right hemisphere), young adults show a large focused activation in the frontal cortex (responsible for doing hard mental work), as well as a focused activation in the parietal area. Older adults show a smaller activation in the frontal cortex, and large, diffuse activations in the temporal/parietal areas. Finally, when studying a picture to make an immediate judgment about it, young adults put the information into long-term storage, activating the hippocampus (above, right). Older adults do not seem to engage this site for a task that does not require long-term-memory processing.

Remembering Names

Why does remembering names get harder as we get older? Name retrieval is both the most common and most embarrassing memory glitch. All of us—whatever our age—sometimes find ourselves fumbling to recall someone's name during an introduction. Perhaps you're shopping with your spouse in a supermarket when you encounter a colleague from work. What should be a simple introduction turns into an embarrassed search through your memory bank for your colleague's name. The philosopher

show what Park calls "a compensatory kind of processing," where both the right and the left frontal cortex are activated.

Park's findings provide an explanation for the complaints frequently voiced by older people of differences in mental "functioning." I hear such complaints on a daily basis in my neuropsychiatric practice. Questioning the person who lodges such complaints reveals not so much a falloff in performance as a "gut feeling" that his or her brain is operating differently from the way it did decades earlier. Park's findings support the validity of such self-observations. Even when the younger person and the older person perform equally well on a given task, their respective brains go about the task in different ways.

Furthermore, and I find this slightly more disturbing, aging seems to involve preferentially those parts of the brain that mediate valued frontal lobe functions that we would least like to lose, such as our ability to organize, look beyond the immediate moment, and entertain several trains of thought simultaneously.

Now the Good News   Does this mean that those of us who are over 40 should despair? That the mature brain holds for us no more surprises, no more pleasures? "Not at all," according to Denise Park. "There is reason for optimism. While it's true that you get slower with aging, that slowing can actually work to your advantage. For one thing, older people are better at mulling over situations, reflecting, and drawing upon their life experiences to arrive at decisions. It's not just a coincidence that the people we have running our country and occupying important positions are not in their 20's, 30's, or even usually their 40's. And I think that's because older people—on the basis of extensive lifetime experience —are wiser. They can put things into context, take a broader view, reach a decision on the basis of less information. That's not to claim that every older person can do these

John Stuart Mill came up with the explanation for such dilemmas. "Proper names are not connotative. They denote individuals who are called by them: But they do not indicate or imply any attributes as belonging to those individuals." In other words, every Joyce in the world could just as easily be named Kate; every John could just as easily be named Robert.

"Names are so out of context," says Denise Park, research scientist at the Center for Aging and Cognition at the University of Michigan, in Ann Arbor. "There's no place where they readily fit into your memory."

Inattention also contributes to many memory failures. For instance, when meeting someone at a cocktail party following an event-filled day at the office, you might be distractedly thinking of office matters at the moment of introduction. With age, the distraction problem increases, thus making memory failures for names especially common.

things. But there's evidence to indicate that life experiences make people much more flexible. Extensive life experiences make a person less likely to make snap judgments or judge others too harshly—in a phrase, to realize that a person or a situation may be something other than what's gleaned from first appearances."

More Good News   And there's some additional good news about the mental abilities of the older person. Language skills do not deteriorate; vocabulary remains steady and can even improve if efforts are directed toward that goal. Such efforts can involve working crossword puzzles or simply taking the time to look up unfamiliar words in the dictionary. I.Q. also remains the same. So, too, does abstract thinking and its verbal expression. Indeed, professional expertise increases with age. For instance, most major law firms are most likely to assign a complicated legal problem to an older, more "experienced" lawyer. The reason is that the older, more experienced practitioner will know, on the basis of accumulated wisdom, how best to proceed.

"Contrary to the popular notion, people don't tend to get worse at their jobs as they age," says Park. "They may process information and react to that information a bit more slowly, but that slight lag in the speed of responsiveness is more than made up for by the knowledge they've accumulated over the years."

Basically, the defining challenge for the aging brain is to somehow maintain healthy circuits and healthy synapses. Especially important are the hippocampus and the nearby entorhinal cortex. Both areas are critical to the formation of new memories and the recall of older ones.

For instance, consider how your brain forms the memory for, say, your breakfast earlier today. That memory includes many things: the smell and taste of the food, the paper you scanned, what you read in that paper, perhaps even

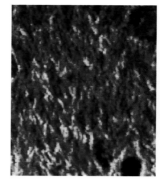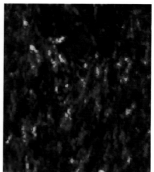

Research shows that receptors for NMDA, a protein important in memory formation in the hippocampus, are more plentiful in young monkeys (near right) than in old monkeys (far right). Because these decreases occur without significant neuronal loss in these critical circuits, scientists theorize that treatments to increase NMDA receptors might help with age-related memory impairment.

a few tidbits of useful information you may have heard while looking up at the *Today Show* or *Good Morning America*. All of these different sights, sounds, smells, and tastes don't exist in isolation within your brain but, instead, form a composite, integrated experience that's etched within your brain's synaptic patterns. Thanks to the synthesizing powers of your brain, you think back to your breakfast as an integrated experience rather than a hodgepodge of different sensory experiences and your conscious responses to them. How does this process take place in the brain?

Establishing a Memory   In essence, neurons from many areas of the brain direct their impulses toward neurons located in the association areas of all of the brain's major cortical areas. Within the association areas, integration takes place. For instance, images of the objects on your breakfast table travel to the brain's visual area, where they are funneled to the visual association area. Sounds travel from the auditory receptive area to the auditory association area. Finally, within the brain's general association areas, the elements of your breakfast experience are integrated into a whole. At this point, the highly processed "breakfast experience" is funneled into the medial temporal lobe for temporary storage and retrieval. If there is any interference with this process, memory isn't established —you don't remember what you had for breakfast, perhaps even whether or not you *had* breakfast.

Forgetting about one's breakfast—indeed forgetting in general—brings to mind, of course, the memory failures characteristic of Alzheimer's disease.

"In Alzheimer's disease one of the first events is the degeneration of the circuit that connects the nearby entorhinal cortex with the hippocampus," according to John Morrison, a neuroscientist at Mt. Sinai Hospital in New York City, studying memory and the effects of aging. "And it's a profound degeneration;

the circuit is essentially destroyed. As a result, the person with Alzheimer's experiences a disconnection between memories for previous events and the ability to lay down new memories."

Although we will discuss Alzheimer's in more detail later, I mention Morrison's comment here to make a specific point: No matter what our age or mental condition, memory is encoded within the brain by means of a specific circuit. If that circuit is altered, various forms of memory impairment can result, ranging from a loss of accuracy and clarity in older, healthy 60- and 70-year-olds to a total inability to form new memories in those afflicted with Alzheimer's disease.

Unfortunately, certain memory impairments seem to be a necessary accompaniment of even healthy aging. Specifically, the rapid retrieval of information is slower in older people compared to younger ones. But in real life—as compared to the psychologist's testing room—such an "impairment" is unlikely to be of much consequence.

Consider once again our nation's former poet laureate. At 95, Stanley Kunitz continues to write and remember poems, performing them in front of audiences with great force and conviction. What's happening in his brain that enables him to successfully do all this?

Remaining Both Plastic and Stable   "The brain of a successfully aging person like Kunitz has to accomplish some extraordinary things," according to John Morrison. "It has to keep neurons alive and in communication with each other. In addition, it has to keep synapses *plastic* so that new things can be learned and *stable* so that those new things aren't forgotten." As Morrison readily admits, "Kunitz may have some genetic help that we don't understand very well. It's my guess that on the microscopic level Kunitz's brain is 'cleaner' than that of most people 20 or more years younger."

Morrison's last point is a sobering one. If we live to 95, all of us would like to imagine ourselves being as clear thinking as Stanley Kunitz. But none of us had anything to do with selecting our genes. In addition, no one so far has come up with a formula that could guarantee the fulfillment of our wish to be as mentally sharp as Kunitz. At the moment, the most promising leads toward developing such a formula involve the synapse and the neurochemicals that influence it.

Is Alzheimer's Disease Hereditary?

So far, five distinct genes—located on chromosomes 21, 19, 14, 1, and 10—are known to predispose a person to Alzheimer's. Chromosome 21 is abnormal in both Alzheimer's disease and Down's syndrome, which is also an illness characterized by distinct physical features and brain abnormalities: plaques, tangles, and memory impairment. Furthermore, as children with Down's syndrome age they develop both the symptoms and brain abnormalities associated with Alzheimer's. Indeed, the presence of amyloid plaques in both illnesses is one reason why researchers believe that Alzheimer's results from the mutation of a gene present on chromosome 21 known to contribute to the encoding for the chemical precursor of amyloid. And if it's true that Alzheimer's develops on the basis of abnormalities in the processing of amyloid, the identification of the other four chromosomes (19,14,1, and 10) suggests that other so far unidentified genes on these chromosomes may also play a role.

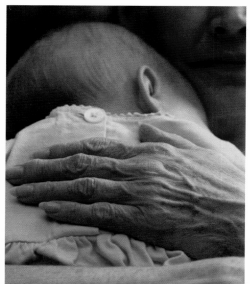

One thing is certain: First-degree relatives of people with Alzheimer's have an increased risk for the disease that approaches 38 percent by age 90. Such figures strongly support the importance of genetic factors, although the mode of transmission is certainly more complicated in most instances than a simple autosomal trait (in which 50 percent of firstdegree relatives develop the disease) or an autosomal recessive trait (in which 25 percent develop the disease).

In one sense, these statistics are somewhat comforting—implying that having a parent with Alzheimer's doesn't necessarily increase the possibility of coming down with the illness. However, Alzheimer's does occasionally pass from one generation to the next by autosomal dominant transmission (from great-grandparent to grandparent to parent). Genetic studies of these familial autosomal dominant (FAD) pedigrees have turned up four distinct FAD genes on chromosomes 21, 14, 1, and 19.

The Alzheimer disease gene on chromosome 19, known as the ApoE gene, is particularly intriguing. Elevations in one variation of this gene (ApoE-4) carry with it an increased risk for Alzheimer's. But this does not by any means imply genetic predestination. Even among people with high ApoE-4 levels the risk remains statistical: The disease cannot be predicted in a particular individual. Thus, as with many genetic mutations, ApoE-4 does not serve as a marker for anything more than an increased risk, a propensity, for the development of disease. Another encouraging statistic is that only about 50 percent of Alzheimer's disease is accounted for by the five FAD genes.

As genetic research progresses, it's possible a reliably predictive test for Alzheimer's disease will become a reality. But such a test will raise ethical questions even more vexing than the scientific ones. Should these tests be available to anyone who asks for them? Should they be available to insurance companies that might want to cut their losses by excluding individuals with a high genetic likelihood of coming down with the disease? Should the results of these tests become part of a person's "permanent record," which can be accessed by the government, employers, and other interested parties that might stand to gain from the information at the expense of the affected person?

So far neuroscientists and our society as a whole have only just begun to wrestle with these vitally important questions.

"The synapse—the point of communication between neurons—seems to be the site of aging, rather than any loss of neurons," says Morrison. "That's because, in contrast to popular opinion, aging isn't accompanied by a significant degree of neuronal death. For this reason, the aging synapse is becoming a major focus for neurobiologists interested in understanding aging and the neurological causes for late-life mental decline."

The Importance of the Synapse   Since the synapse is likely to play a major role in enhancing mental performance in later life, several important points about it should be kept in mind. Consider the following: The brain is totally reliant for its functioning on synaptic communication involving trillions of synapses. In turn, each neuron receives thousands of synapses. When these synapses are functioning normally, information in the form of neurotransmitters flows across the synaptic gap and links up with specific receptors.

In addition, all of the medications that work on the brain do so by modifying some aspect of synaptic functioning. For example, as we have seen, all successful antidepressants work by increasing the concentration of one or more of the neurotransmitters within the synapses. In addition, hundreds of proteins and genes within neurons—all of them potentially modified by aging—influence synapses and thereby also influence the communication among neurons. Memory, for instance, can be usefully thought of in terms of synaptic functioning. "Whenever you remember an event it's because at the time of the event certain molecular processes occurred that strengthened synaptic connections involving many neurons," says Morrison.

When neuroscientists study the synapse they come face-to-face with the basically communal nature of the brain. Within the brain a single synapse—like a single neuron—is meaningless in the grand scheme of things. But an abundance of synapses, each interacting with different neurons, establishes a vast network of communication. As we saw earlier, the infant brain is charged with constructing that network; the task of the mature brain is to maintain it. And that task is reliant on interactive events at the level of the individual neuron. Of course these cellular and chemical events can go awry. The most dramatic and frightening example is Alzheimer's disease.

Alzheimer's, the Most Common Cause of Dementia   First described in 1907 by neuropsychiatrist Alois Alzheimer, the disease that bears his name was initially considered a rare disorder affecting only people under the age of 65. Today it is recognized as the most common cause of what was once called senility and is now known as dementia. Alone it accounts for 50 percent of all dementias, with an additional 15 to 20 percent resulting from a combination of Alzheimer's disease and vascular disease of the brain.

Starting in the fourth decade (when it's extremely rare) the prevalence of Alzheimer's disease increases with each succeeding decade. At least one person in four over the age of 85 is afflicted. But such dry statistics camouflage the human tragedy of a deadly illness that so far has resisted scientists' best efforts toward finding a cure.

Here is an example of the devastation wrought by the disease. Imagine yourself as a neurologist. Picture a well-dressed, pleasant, indeed charming woman of 55 who sits in your office one spring afternoon and tells you about her increasing difficulties with memory. She speaks of her recent retirement from her medical practice because of several incidents when she momentarily confused one of her patients with another. While you're talking to her nothing seems particularly out of sync. You don't observe any problems in her speech, mood, or thinking processes except perhaps for a mild memory impairment that comes to light only after you administer standard memory tests. But the memory loss isn't extensive enough on this initial visit to enable you to make a confident diagnosis.

Now fast-forward your mental video four years and picture that same woman sitting in your office once again. Although still neatly attired, she can no longer select her clothes and has to be helped by a caretaker-companion to get dressed for her trip to the office. This time she didn't drive but was brought to your office by her husband. She has no memory of having seen you regularly over the past four years, nor of any of the medications you administered to her over that period in an unsuccessful effort to hold in check what has turned out to be a devastating illness. She doesn't know the day or the date, cannot remember the names of three objects (lemon, key, ball) for more than a few seconds, and, unless verbally prodded, will simply stare into space.

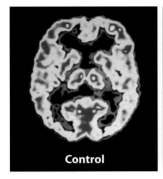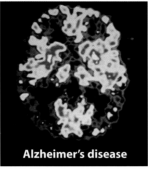

Control **Alzheimer's disease**

PET scans show differences in brain activity (measured as glucose metabolism) between a normal brain and a brain affected by Alzheimer's disease. Blue and black denote inactive areas. Researchers are uncertain whether the decline in glucose metabolism in Alzheimer's causes neurons to degenerate or whether neuron degeneration causes metabolism to decline.

Regrettably, this unfortunate woman isn't taken from a work of fiction, nor is she a composite. She is one of my own patients afflicted with Alzheimer's disease. Here is her husband's account of a typical day:

"Once an avid reader of books, she no longer even looks at the newspaper. She sits for hours staring at the television, but she can't tell me what she's watching. She can't adjust to any change in her daily schedule and, at night, easily becomes lost in the house that we've lived in for more than 30 years."

Other relatives of Alzheimer's patients we talked to told similar stories of gradual losses in mental capacities. Here is how Karen Johnson described the process:

"My husband, who loved to cook, gradually began to lose his ability to shop for the food items he needed for a recipe. This progressed to the point where he couldn't get three or four things together all at the same time. The rice would be ready, but the vegetables were still in the refrigerator and the meat was still marinating. He couldn't figure out how to get everything done at the right pace and in the correct order. But the worst problem surfaced on the day I got up to make oatmeal for breakfast, opened the cupboard where the cups and pans were kept, and found a large bag of spinach he had placed there. At that moment I realized he had lost the ability to figure out the appropriate place for things."

As with my patient and Karen Johnson's husband, the initial impairment in Alzheimer's disease usually involves memory—not the kinds of memory problems we all experience from time to time (Now where did I park my car an hour ago in this crowded shopping center parking lot?), but rather what Denise Park refers to as a "qualitative change" (Did I drive to this shopping center, or did I walk here from home?).

Qualitative Changes   "In normal aging you might forget the name of the restaurant that you went to yesterday. But in Alzheimer's you forget that you even went to a restaurant or the name of the person you had lunch with. Nor does a normally aging person forget how to use a checkbook or prepare a meal. If you have Alzheimer's, you can get lost while driving from your daughter's house to your own home, a journey you've made hundreds of times. In the morning you can't figure out how to put together a coherent outfit for the day and as a result simply stand there staring into your closet. Such impairments, frequently observed in Alzheimer's disease, represent qualitative changes—a mental cliff from which the person with Alzheimer's has fallen."

Park's comments should be reassuring to those who fear that a momentary memory lapse may portend the onset of Alzheimer's disease. Some degree of forgetting is a perfectly normal accompaniment to healthy aging. In addition, it's useful to remember that memory powers differ greatly among people at every age. Memory lapses are also explained differently according to the age of the person with the "memory failure." A teenager who can't come up with someone's name is simply considered to have a "poor memory" or not be "paying attention." An equal degree of impairment in a middle-aged or older person, however, is likely to raise questions about possible Alzheimer's disease. But even among the aged, reassurance is perfectly appropriate in most instances when a mildly lapsing memory is the only sign of a mental faculty affected by aging.

"Alzheimer's is really a disease of cognition—the ability of the brain to attend, identify, and act on complex stimuli," according to Mt. Sinai's John Morrison. "And cognition doesn't arise from a single area of the cortex but results from the cohesive interaction of many parts of the cortex. Thus, in Alzheimer's you find a loss of connecting neurons responsible for complicated functions requiring the integrated activity of the frontal, parietal, and temporal cortices."

Depending on the area of the brain most affected, a person with Alzheimer's disease may experience different difficulties. If the frontal lobe bears the brunt of the damage, planning and the integration of complex activities may suffer, as with Karen Johnson's husband, who gradually lost his ability to coordinate the preparation of a meal. If the parietal lobe is selectively affected, the person may easily get lost in familiar surroundings as a result of the death of neurons in those parietal areas responsible for geographic orientation. And since the neuronal losses may differ from one Alzheimer's patient to another, there is no such thing as a "typical case" of Alzheimer's; each one is different.

Selective Vulnerability   As Morrison puts it, "Alzheimer's disease is one of selective vulnerability, with some nerve cells and circuits profoundly affected, while others are never affected. Unfortunately, the most affected cells in Alzheimer's disease are those responsible for functions that are uniquely well developed in humans, such as those residing in the cerebral cortex."

Variation in selective vulnerability explains why Karen Johnson's husband, although he can no longer cook a meal, retains the ability to paint. "My husband's art is still highly developed. It's as if the right side of his brain is still engaged; you can see it in the way he can express himself through his paintings. His painting is the only area where he is still anywhere close to having the brain of a normal 65-year-old."

As another example of selective vulnerability, consider a woman I met socially a week or so ago. Her husband told me his wife had recently been diagnosed with Alzheimer's disease. He spoke (in a muted voice) of her impatience, her tendency to "fly off the handle," and most of all her habit of endlessly asking the same questions but failing to keep track of the answers. When introduced to me, she appeared somewhat distracted but her social responses were perfectly appropriate. Later, while entering the dining room of the restaurant, she sat down at the piano and played a perfect version of "Für Elise." When finished, she played the piece again and would have played it for the third time if her husband hadn't gently led her to our table. Thanks to the selectivity of her brain involvement, her musical abilities remained intact despite her inability to remember that she had just played the same musical composition for the second time.

Given that Alzheimer's is such a serious illness, some people question the wisdom of doctors telling patients and their families of a suspected diagnosis in the early stages of the illness. After all, nobody likes to hear bad news. John Morris, director of the Memory and Aging Project at Washington University in St. Louis, thinks this unwillingness on the part of doctors to make an early diagnosis is ill advised. "A person in the early stages of the illness is still at a stage when he can

Plaques: Cause, Not Effect

For decades scientists believed that so-called plaques and tangles, the postmortem evidence of a brain ravaged by Alzheimer's, were caused by the disease. (In the image at right, plaques are the reddish structures; cells full of tangles are black.) But more recent investigations suggest just the opposite: The formation of plaques and tangles probably causes the degenerative symptoms of Alzheimer's, not the other way around.

Tangles occur inside of brain cells, while plaques are gummy globs that attach themselves to the outside of cells (see next page). The globs are molecules of a protein called beta-amyloid that floats in the space between cells. Most people can survive well into old age with only minimal buildup of beta-amyloid. But in those with a genetic predisposition to it, the amyloid buildup occurs more quickly. The sticky protein binds to both neurons and supporting glial cells, causing an inflammatory response that ultimately destroys neurons. The result is the onset of Alzheimer's disease as early as the 40's or 50's.

A number of research efforts are aimed at finding ways to halt or reverse the buildup of beta-

make choices about his future. He can decide about financial and legal matters and select the person he wants to make future health care decisions for him should he become incompetent." Equally important, drugs are now available that help preserve mental function longer.

"Most people when they hear the term 'Alzheimer's disease' get a mental image of an incapacitated person almost ready for a nursing home," according to Morris. "But actually the milder cases are much more common. People don't change overnight from a healthy functioning adult to a nursing home patient. Instead, the change occurs gradually. That means that we have to diagnosis it in the early stages if we're going to make a difference. At the moment, we have drugs that help maintain the person's abilities. But we're really aiming at the development of drugs that will arrest it. So the earlier we're able to detect it, the better it is for the mildly affected patient."

Diagnosing Early Stages   How does the doctor diagnose Alzheimer's disease in its earliest stages? Although each patient's course may differ, certain distinct features characterize the disease. As already noted, memory impairment is usually the first sign. Next comes difficulty in holding complex discussions or carrying out such everyday activities as balancing a checkbook or coordinating the preparation of a meal. Eventually the basic personality is altered. This can run the gamut from a general apathy to outbursts of temper, even physical assaults.

amyloid. One approach that has shown promise in mice is a vaccine that causes the body's immune system to produce antibodies that mark beta-amyloid for attack by the brain's immune cells. Mice given the vaccine have shown significant reduction in plaque formation.

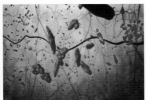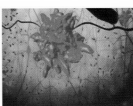

Adrift in the spaces between brain cells are thousands of molecules of a protein called beta-amyloid. Over many decades, the molecules may clump together to form sticky globs that attach themselves like barnacles to the outside of cells. These clumps disrupt the normal function of the neuron, setting off a chain of events that destroys the neuron itself—bringing on the impairments of thinking, memory, and reasoning that characterize Alzheimer's disease.

"In a nutshell, Alzheimer's disease involves the brain areas important in thinking, memory, and reasoning," says John Morris. "As a result, the person becomes increasingly dependent on other people. Eventually the person needs help in carrying out basic functions like dressing, showering, shaving, and brushing one's teeth. Novel innovative activities are relinquished in favor of simpler ones. For instance, an experienced cook may revert to simpler, comparatively undemanding recipes. The key element here is *change*: a loss or deterioration in previously established skills. This, of course, will vary from person to person. For instance, since I've never been particularly handy around the house, the fact that I can't be much help in the remodeling of my kitchen doesn't mean anything. But for someone who has demonstrated such skills earlier in his adult life, the loss of these skills may signal the onset of Alzheimer's disease."

With this as background, the logical question arises: What is the nature of the destructive process that takes place in the brain of a person afflicted with Alzheimer's disease?

Accumulated Debris   Put simply, the Alzheimer's brain accumulates debris—referred to as plaques and tangles—that build up like discarded items crammed into an already overcrowded attic. When a section from such a brain is observed under a microscope, the tangles look like twisted railroad tracks, paired helical filaments formed from a molecule called tau and located *inside* the neurons. Plaques, in contrast, appear as unprepossessing gummy globs of spherical material *outside* the neurons and consisting of a central dark core surrounded by a darker rim composed of degenerating neuronal processes.

As the debris increases the accumulation eventually disrupts nerve cell communication. Consequent to this loss of communication, neurons start to shrink, die, and finally disappear. When the clutter reaches a critical point (varying from person to person), the resulting loss of brain cells leads to those disturbances of thinking and behavior that characterize the illness.

While neuroscientists have observed plaque accumulation for decades (dating to Alzheimer's original patient seen in 1907), the discovery of its chemical composition is comparatively recent, the key breakthrough occurring in 1984. In that year, neuroscientists learned that the core component of plaque consists of a protein called beta-amyloid. Because this protein is greatly increased in people with Alzheimer's disease, many researchers are now convinced that beta-amyloid is the key to understanding and treating Alzheimer's disease.

Gradual Amyloid Buildup   We asked Dennis Selkoe, a neuroscientist affiliated with Harvard Medical School and Brigham and Women's Hospital, why Alzheimer's primarily happens to old people. "Because time is of the essence," Selkoe explained. "The buildup of amyloid is a gradual process that takes decades. All brains have some amyloid in them; it's an inevitable consequence of getting older. But only some people accumulate enough amyloid to result in Alzheimer's disease. Most people can survive into their 60's, 70's, and 80's with only minimal buildup. But in those people with a genetic predisposition—perhaps a grandparent or a parent with Alzheimer's —the amyloid buildup occurs more quickly and the person comes down with the illness at the same age or even an earlier age than their relative. In my view, if we could stop that buildup of amyloid, we wouldn't see Alzheimer's disease."

One finding overridingly favors the beta-amyloid hypothesis: Beta-amyloid is elevated in those cases of Alzheimer's that can be traced to defective genes. "The number one support for the amyloid hypothesis is that all of the genes that cause Alzheimer's disease have also been shown to crank up amyloid early in the course of the illness," says Dennis Selkoe. "And you don't have to do fancy tests to show that. You can find this excess amyloid not only in the brain but also in cells from the skin or blood."

"In other words these defective genes are a smoking gun pointing toward beta-amyloid," according to neuroscientist Dale Schenk of Elan Pharmaceuticals, Inc. "Thus, many of the future therapies we're trying to develop now are based on either blocking beta-amyloid from ever getting made in the first place or getting rid of it once it's there in the brain."

Once in the brain, amyloid sets off a cascade of reactions. Since amyloid is a sticky protein, it binds to itself as well as to the surfaces of neurons and the supporting glial cells. Furthermore, it stimulates the glia to pour out proteins involved in inflammatory responses.

"Inflammation in the brain is something you'd rather not have," says Dennis Selkoe. "Even though the inflammatory response that occurs around the amyloid plaques represents the brain's defense against amyloid, inflammation only causes even more trouble. For example, when the brain tries to defend itself against amyloid, its effort leads to the release of a lot of chemicals that cause even more of a mess. It's sort of like what happens when somebody tries to break up a fight between two individuals. The person ends up being drawn into the fight himself and ends up with his own black eye. Ultimately the efforts weren't worth it."

In support of Selkoe's point, several recent studies suggest that people who regularly take anti-inflammatory drugs like ibuprofen (Advil) may be slightly less likely to develop Alzheimer's. So should everyone rush out and buy anti-inflammatory drugs? "We still don't have enough hard clinical proof that they work," says Selkoe. "But there's something there—enough smoke to suggest some fire."

Vaccination? A vaccine may provide another promising, less-injurious approach to getting rid of beta-amyloid. Dale Schenk has administered a vaccine to mice, with surprising results. His purpose was to produce antibodies that would flag beta-amyloid for subsequent attack by the body's immune cells. This is not an entirely new concept, of course. All commonly employed vaccines operate on this principle of creating an immune response against a bacterium or virus that has gained entry to the body. But a vaccine against Alzheimer's disease would have to differ from the usual vaccine in an important way: Beta-amyloid is not a foreign invader but a substance made within the brain.

"We had the idea that if we created a lot of antibodies by the use of a vaccine, a few of them might get into the brain and orchestrate an attack against the beta-amyloid and either destroy it or alter the way it forms the plaque. This, in turn, might alter or reduce the brain tissue damage associated with Alzheimer's disease. Essentially, what we're doing with the vaccine is tricking the immune system into acting as if the plaque is a foreign substance that must be destroyed."

So far, the vaccine has only been applied to animals. But the results are particularly impressive in a genetically engineered mouse that contains a defective human gene associated with Alzheimer's. This defective gene leads to the development within the mouse brain of the plaques associated with Alzheimer's. Administration of the vaccine to the mouse leads to a reduction in the number of plaques. This holds true both for mice that already have plaques and those mice in which the plaque formation hasn't yet begun.

According to Schenk, "The vaccine stimulated the immune system to go in and clear out the plaques in the brain tissue of these mice. Not only did we get rid of or reduce the plaques, but the animals performed better on standard tests. Our results were beyond anything we had hoped for. We hope that this result will occur in human patients, because if it does we'll have a real treatment for this disease."

At this point, however, vaccine administration to humans exists as only a gleam in the eye of researchers. First, the vaccines must be shown to be safe. If safety doesn't prove a problem—and that may turn out to be a big "if"—clinical trials can be started in human subjects.

Another promising future treatment aims at suspending the chemical reaction leading to the formation of amyloid. Ordinarily, enzymes known as secretases act like molecular scissors that cleave healthy, normal proteins, amyloid precursor proteins (APPs), into the neuron-suffocating amyloid plaques. Treatment with secretase inhibitors keeps the scissors from closing by blocking the enzymes so that the APPs aren't cleaved into molecules of amyloid. Several pharmaceutical firms are currently involved in early human testing of secretase inhibitors. But the trials face one monumental and potentially limiting impediment: Scientists can't be certain if blocking secretase enzymes—which have normal functions in the brain—could boomerang and lead to a reduced ability to combat other diseases. Until such issues are resolved, both the secretase inhibitors and vaccines remain promising but still unproven approaches to the treatment or even prevention of Alzheimer's disease.

Besides the amyloid plaque, the dying neurons in Alzheimer's disease also contain tangles of tau, a neurofilament that gives the cell its structure and stability like the structural beams within a building. But as the tau filaments entangle, the structural underpinnings of the cell are weakened. An alternative approach to Alzheimer's disease concentrates on the tangles. Currently, this is a much less popular approach because the tangles only occur in the dying neurons after the accumulation of amyloid.

"If you choose to treat the tangles, which are really end-stage lesions, you're trying to come in at the last minute rather than early in your rescue efforts," says Rudolph Tanzi, director of the Genetics and Aging Unit of Massachusetts General Hospital, in Boston.

While neuroscientists continue to sort out the tangle–amyloid debate, patients, families, and doctors are left with partial and unsatisfying treatments. As already mentioned, anti-inflammatory drugs like Advil may dampen the inflammatory reaction related to the brain's response to amyloid accumulation. Another nonprescription drug approach aims at decreasing the effects of a form of oxygen that damages nerve cells.

Oxygen is essential to sustain life, but a tiny fraction of oxygen molecules form free radicals: fragments of molecules or atoms containing an unpaired electron. Since unpaired electrons are unstable, the molecules or atoms with

Inside the brain cell are long chains of molecules that transport nutrients from one end of the cell to the other. Keeping those chains stable, much as railroad ties stabilize the rails of a track, are microtubules made of a protein called tau. In Alzheimer's disease, tau filaments rupture, the strands of the transport molecules separate and collapse, and the cell dies, choked by tangles of tau.

The Importance of Exercise

When the jogging craze struck in the mid-1970s, thousands of people suddenly discovered the almost addictive properties of exercise. Getting the blood going and working up a sweat seemed to have a beneficial effect on not only the muscles but also the mind, easing tension and producing a sense of well-being.

Now neuroscientists have found that exercise has a direct effect on the health of brain cells, in particular stimulating the production of a growth-inducing protein called BDNF. Active exercisers in their 90's, such as Red Simmons (right) and Milton Adamson (opposite), can attest to the long-term benefits, both physical and mental, of regular vigorous exercise.

unpaired electrons attempt to achieve stability by stealing electrons from vital molecules, such as those that form the outer membrane of neurons. The free radicals also attack the proteins that make up the neuron's internal structural components—such as the DNA molecules in the nucleus of the neuron. In an effort to prevent this free-radical damage, neuroscientists suggest use of the antioxidant vitamin E. This easily obtainable vitamin counters the negative effects of free-radicals by soaking them up and thereby slowing the destructive electron-grabbing chain reaction.

Changing Lifestyle   Additional help in warding off Alzheimer's disease may also come from a series of lifestyle changes such as exercise to increase blood flow to the brain. As an example of the benefits of exercise, consider Milton Adamson, a 94-year-old, who has exercised daily since he was 42. On memory tests Adamson scores as well as people half his age.

Adamson's mental performance has been tested at the University of California at Irvine, where neuroscientists like Carl Cotman, director of the Institute for Brain Aging and Dementia, are searching for the biochemical keys to successful aging. To evaluate whether exercise works by stimulating proteins that keep neurons healthy, Cotman compared two groups of rats. While one group sat idly in their cages, the other rats spent hours a day exercising on running wheels. After eight days Cotman found that in the exercising rats levels of a growth-inducing protein, BDNF, had doubled.

A homicide detective during Prohibition, Red Simmons (opposite) retired at age 46, then began a second career as a track coach before retiring from that at age 84—about seven years ago. Like Simmons, Dr. Milton Adamson (left) looks decades younger than his 94 years and scores as well as people half his age on memory tests. Adamson swears by the regimen of daily exercise he began when he was in his early 40's.

"The surprising bottom line of the research is that exercise will stimulate the production of all kinds of wonderful molecules that keep neurons healthier and stronger," Cotman says. "I think of them as molecular fertilizer. And exercise actually increases these molecules in the brain."

Other lifestyle factors thought to be important to longevity and decreasing the chances of developing Alzheimer's disease and other dementias include curiosity; keeping busy; reducing stress; getting adequate sleep; seeking diversity and novelty; keeping up friendships and social networks; thinking of education as a lifetime process rather than something limited to life's early years; and finally, health and dietary changes aimed at maintaining normal blood sugar, cholesterol, and blood pressure. Such general lifestyle changes are especially important in those people who either already have mild Alzheimer's or are at increased risk of it.

"If you know you have a tendency to develop Alzheimer's disease, you want to do everything to avoid a double whammy from any other brain problem such as a stroke," according to Dennis Selkoe.

Once a person has the disease, unfortunately, currently available medications provide only temporary help by increasing the amount of acetylcholine—the principal neurotransmitter decreased in Alzheimer's disease. Ordinarily acetylcholine, after release into the synaptic cleft and linkage with its receptor, is released from the receptor back into the synapse for eventual destruction by an enzyme, acetylcholinesterase. In contrast to normally aging people, the Alzheimer's patient has decreased amounts of acetylcholine in his or her brain. This is especially true in the hippocampus and the nucleus basalis of Meynert, a collection of neurons deep in the brain that give rise to most of the brain's acetylcholine-transmitting cells.

Current treatments with drugs such as tacrine (Cognex), donepezil (Aricept), and rivastigmine (Excelon) are directed at increasing the amount of acetylcholine in the synapse by inhibiting its normal breakdown by acetylcholinesterase. These acetylcholinesterase-inhibiting drugs provide only temporary benefit.

"A drug like Aricept is better than nothing," according to Rudolph Tanzi. "It helps an Alzheimer's patient to make the best use of the nerve cells that are still surviving in the brain. Unfortunately, after a certain amount of time, the effects can start to wear off. Even though it's useful, Aricept is still just a Band-Aid on a gushing wound. But right now we just don't have any other therapy."

The Thunderclap of Stroke   While dementia—and normal aging, as well—develop slowly and insidiously, stroke happens with the speed of a thunderclap. Indeed, Hippocrates wrote of stroke as *plesso* meaning "to be thunderstruck."

A stroke results when blood flow is interrupted by a blood clot or ruptured artery serving the brain. As the result, millions of neurons are killed and millions more are injured. Following a stroke, the damaged part of the brain shrinks to half its former size as scar tissue takes the place once occupied by healthy neurons.

Eighty percent of strokes are ischemic, resulting from closure by a blood clot of a major artery supplying the brain. The clot may result from either a buildup over many years of fatty deposits on the walls of an artery serving the brain (a thrombotic stroke) or from a clot formed elsewhere in the body that breaks loose and wends its way to the brain (embolic stroke). The remaining 20 percent of strokes are hemorrhagic, which means they result from rupture of an artery within (intracerebral) or around (subarachnoid) the brain. Strokes occur most often in older people, but the illness is no respecter of age. Intracerebral hemorrhage can happen even before birth and result in cerebral palsy or other serious brain impairments. Subarachnoid hemorrhage can happen at any age since it's based on an inherited weakness of the artery's wall, which renders it more likely to rupture under conditions of elevated or even normal blood pressure.

Stroke Patterns Strokes take many patterns, depending on the areas of the brain damaged by the interruption of blood supply. In many cases neurologists can diagnose the exact blood vessel occluded or ruptured on the basis of their knowledge of the brain's pattern of blood supply. They do this by observing certain constellations of symptoms (complaints) and signs (objective indications) that serve as markers (see next page).

Traditionally, most doctors believed—and informed their patients and the patient's relatives—that any recovery from a stroke would occur within the first few months. After that period, the doctor held out little hope for recovery. With the entrenchment of this pessimistic outlook in doctors' minds, their patients received scant encouragement to continue their rehabilitation efforts; they gave up any expectation of getting better and simply stopped trying. New hope for stroke victims emerged from experimental observations dating from the 1980s on paralyzed monkeys.

After suffering severance of the nerves that send sensory information from an arm to the brain, monkeys can no longer feel that arm. Since the motor nerves are left unsevered, the monkeys should be able to move the arm, but they can't; it hangs limply at the monkey's side. But full functioning of the paralyzed arm will return if the monkey is placed for up to two weeks in a restraining device that prevents the monkey from using the normal arm. Later, after release of the normal arm from the restraint, the monkey will continue to use the no-longer-paralyzed arm in concert with the other arm. This so-called constraint-induced (CI) movement therapy is being successfully applied to humans paralyzed as a result of strokes.

"I had the idea that the same techniques that proved successful with monkeys would also be effective in enabling people to use their paralyzed arms after strokes," according to Edward Taub, a specialist in psychology and rehabilitation at the University of Alabama at Birmingham. "After a few months following the stroke the person has *learned* not to try to use the paralyzed arm or leg. This is a conditioned response that, as time goes on, becomes increasingly powerful.

"But the adult human brain is enormously plastic. The amount of brain area involved in producing movements or receiving sensation keeps changing continuously based on the amount of use of that part of the body. In a sense, the brain

The Sites of Stroke

By assessing a patient's symptoms and signs, neurologists can determine which part of the brain has been damaged. The most common symptoms and signs of stroke based on the stroke's location are:

Right hemisphere:
- Left-sided weakness or paralysis
- Denial or indifference to the paralysis
- Loss of sensation on the paralyzed side
- Loss of vision in both eyes for objects off to the left
- Confusion
- Disorientation for time and location
- Emotional instability
- Dulled responsiveness
- Poor judgment
- Impaired ability for rational thought

Left hemisphere:
- Right-sided paralysis
- Loss of vision in both eyes for objects off to the right
- Problems in speaking and/or understanding the communications of others
- Depression, slowness
- Impaired thinking
- Temporary confusion

Cerebellum:
- Loss of balance
- Dizziness
- Nausea, vomiting
- Flaccid, "rag doll" weakness of the arm and leg on the damaged side

is like a muscle and the more that you exercise it, the better the brain becomes at carrying out its respective functions."

In order to exercise those parts of the brain affected by the stroke, Taub places the patient's normally functioning arm in a restraint that resembles a baseball catcher's mitt. The mitt is then placed in a sling hanging from the neck and closely fitted to the chest. "We're forcing the patient to overcome the learned tendency not to use the weaker arm. With the normal arm restrained, the patient doesn't have a choice. He has to use the weaker arm."

Over the next 12 days for six hours each day, the patient tries using the paralyzed arm in repeated exercises ranging from stretching the arm out from the body to picking up small objects or spooning food. At first, the efforts are clumsy and only partially successful. But with practice the movements improve.

Repetition is the Key In CI movement therapy the patients have to repeat movements over and over again. Repetition is an important part of the therapy. Indeed, it's this repetition and practice of the movements that produces the desired change in the brain's organization and functioning. "We're finding that the brain is a muscle, and the more you exercise it the stronger it gets," according to Taub. As proof of his assertion, he points to a prize patient, James Faust, of Calera, Alabama.

A stroke paralyzed Faust's right arm so completely that Faust even considered approaching a surgeon with the request that the right arm—"that was just danglin' "—be amputated. But only a few weeks after starting Taub's program, Faust and his wife saw a dramatic change.

"One evening when we were out to dinner I looked over at my wife and her eyes were wide and her mouth was agape. I said 'What's the matter?' And she said 'Look at your right hand.' My right hand was holding a steak knife and I was cutting my steak as if nothing had ever happened to that hand. And from that day on, I've really started using my right hand." Encouraged by his progress, Faust has continued with his exercises, and today he has "no problem doing anything I want to do with my right arm. I can drive my car, tie my shoes, brush my teeth, and shave."

During our visit with Faust, we watched him skillfully clipping the hedges in his backyard. His use of his right hand was indistinguishable from that of any person his age who had never suffered a stroke.

To discover the basis for improvements like Faust's, Taub employed a focal magnetic stimulator that detects the areas of the brain where various muscles are represented. He concentrated on the brain area responsible for the small muscle that moves the thumb and used it as a surrogate for the paralyzed arm. Measurement prior to CI movement therapy showed 12 active positions on the magnetic map (a normal thumb shows up to 30 active positions). Mapping carried out after completion of the CI training showed an average of 22 active positions on the map, nearly double the pretherapy number.

Neuron Recruits   Taub believes that the focal magnetic stimulator findings are based on the increased recruitment and activation of neurons surrounding the area of primary damage caused by a stroke. "Neurons that haven't been killed by the stroke but are in the vicinity of the damage behave like plants and send out elements that make connections with other neurons. There are also probably neurons involved in what we call 'silent connections.' These are connections that were not previously operative but become active after the stroke. We believe that this activation and recruitment of neurons increases as the patient engages in repetitions of the newly learned movements."

With his good arm in a restraining mitt, a stroke patient performs repetitive tasks designed to exercise the parts of his brain affected by the stroke. Constraint-induced movement therapy, as it is called, forces the patient to overcome the learned tendency not to use the weaker arm.

Taub is careful to add that nobody is sure why CI movement training works. While rehabilitation specialists like Taub know that the size of the brain area involved in movement of the paralyzed arm increases, they aren't sure of the mechanism. But the essential process involves a rewiring of the brain. And that rewiring is possible only because the adult brain retains a good part of the plasticity it had when it was very immature. No matter how old the person becomes, that plasticity remains undiminished.

One proof of the brain's retained plasticity is the fact that CI movement therapy works even in patients who had their strokes decades earlier and have had limited use of their limb ever since. Here is how Taub's patient Ellen Justice, who had a stroke, described her recovery of the use of her left arm:

"After my stroke I wasn't aware, in a way, of the existence of my left arm. In my daily life I didn't use it or even consider using it most of the time. And on those few occasions when I tried to use it, the result was a disaster. So I just tended to forget about it. But during two weeks of treatment, I was forced to remember that I have a left arm and hand. I was able to get my brain working normally again, recognizing my left arm and using it. Even though I'm not yet able to use a computer in my work as an auditor, I've got the use of two of my five fingers back. But I'm going to have to continue to do the exercises that they gave me in order for me to see any additional progress. At three years after my stroke I've come a long way and I fully expect to make additional gains in the future."

On the basis of successes like CI movement therapy, neuroscientists are eagerly seeking more effective therapies and, at some point, even cures for both stroke and dementia. "And it's important to make a distinction between a cure and a therapy," according to Fred Gage, an investigator at the Salk Institute. "A therapy only addresses some of the key symptoms of the disease. A cure actually

Working around the yard with full use of both arms, stroke patient James Faust of Alabama is a walking testament to the success of constraint-induced movement therapy—and to the brain's plasticity.

protects against some of the changes responsible for the disease or reverses the effects of the disease and restores the healthy state. We're talking here of the distinction between rational therapies based on what we know is going on with the disease versus a cure that basically eliminates the disease."

Adult Neurogenesis   Gage is a pioneer in the field of adult neurogenesis: the production of new neurons in the mature brain. If new nerve cells can be produced, neuroscientists stand a good chance of reversing the effects of stroke, dementia, and other degenerative brain diseases.

Traditionally, neuroscientists believed that new neurons didn't arise in the adult brain. The reasoning went like this: The brain is a very complex structure composed of circuits involved in learning, memory, and storing large amounts of information. The addition of new neurons into these circuits can be expected to disrupt brain function because the new neurons can't be accommodated into the existing circuits.

Neuroscientists were, however, comparing the brain to a computer. Although memory capacity can be increased in a computer by adding a hard drive with greater memory capacity, simply tossing in stray bits of wiring can't do the same thing. But the brain doesn't operate like a computer. For one thing, the organ changes from day to day according to a person's experience and activities. The brain's plasticity is, in fact, its most important operational principle. Neurons alter their connections, make new connections, reorganize, and integrate connections in response to the environment. And if that environment is enriched, the brain is enriched as well.

For instance, adult mice living under the enriched conditions of larger cages, shared company of other mice, and opportunity to play with a variety of toys

The Promise of Stem Cells

One of the characteristics of the embryonic brain is its incredible neuronal production: Hundreds of different kinds of neurons (right) are produced from the unspecialized cells known as stem cells. For decades scientists believed there was no stem cell activity in the adult brain even though stem cells continue their regenerative activity in skin, liver, the lining of the intestines, and bone marrow. Recently, however, researchers have discovered stem cells in the hippocampus of the adult brain. In mouse studies, they have also found them near the brain's fluid-filled spaces called ventricles. It may be possible to induce these precursor cells both to become active and to turn into neurons appropriate to different parts of the brain and central nervous system. This holds out the promise of alleviating, if not curing brain illnesses and injuries.

ended up with slightly heavier brains, differences in neurotransmitter levels, more connections between neurons, and increased neuronal branching. Moreover, the mice performed better on learning tasks like learning to successfully navigate mazes.

So if the brain doesn't operate like a computer, it might just be possible for functional improvement to result from the addition of new brain cells. Furthermore, jettisoning the computer analogy also raises the possibility that new brain cells might very well be incorporated into circuits. But prior to November 1998, no one had ever shown that the mature human brain is capable of spawning new neurons. That month Gage's team at the Salk Institute, along with another team led by Peter S. Eriksson, of the Sahlgrenska University Hospital, in Gothenburg, Sweden, announced the discovery within the hippocampus of stem cells, a versatile cell with an activity level similar to the cells present in the embryo.

Morphing Stem Cells   Stem cells are primitive, unspecialized cells that can morph into mature cells of the body while continuing to multiply and provide a constant source of new cells. Think back to the last time you suffered a cut or broke a bone. Stem cells in the skin and bone account for the healing that took place. Stem cells also give birth to the cells constituting liver, the lining of the intestines, and all the cellular constituents of the blood. Thus, the discovery of stem cells within the brain made it likely that stem cells are capable of turning into adult neurons.

"Stem cells can divide, move around in the brain, end up in specific locations, and then fully differentiate into neurons," according to Gage. "It isn't as if a mature brain cell is undergoing cell division but rather that the process of development is

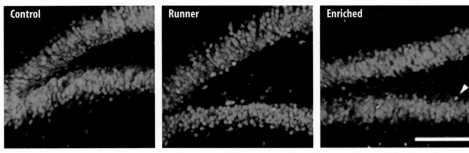

Experiments with adult mice strongly suggest that exercise (running) and an enriched environment (a larger cage and lots of toys) encourage stem cells in the hippocampus to continue dividing and forming new neurons. In one experiment, hippocampal cells were labeled with a chemical marker that becomes integrated into the DNA of dividing cells (red). Four weeks later, they were also labeled to mark neurons (green) and glial cells (blue). As seen above, both the running mice and the enriched mice show many cells that are dividing (red) and others that have differentiated into neurons within the four week period (orange).

continuing in the adult brain. And that's really a key feature in understanding the larger picture of what neurogenesis and plasticity are all about."

While most of the brain stem cells found so far have been in the hippocampus, they are not restricted to this site. Gage uses the analogy of two identical twins raised in different environments. In such instances, it's the environment working on the genetic endowment that accounts for any of the differences observed between the two people. Similarly, stem cells take their cues from the milieu in which they find themselves. This dependence on the surroundings leads to an important principle: "You can't just put stem cells somewhere in the brain and expect them to take over the function of missing or damaged cells. You have to figure out the requirements for turning these cells into neurons."

In the laboratory, Gage places stem cells in a dish of nutrients and by providing the correct chemical stimulation "educates" the cells to become fully mature, chemically active, electrically responsive neurons. "We train the stem cells in a culture dish and expose them to appropriate chemical stimulants, like growth factors, before we transplant them into the brain at a more mature stage. We think of it as sending them to high school, then college, and then graduate school before we put them back into the brain to do their job. Right now we're trying to figure out the training regimens required so that when we put them into a damaged brain area they will turn into the kind of cell that we want."

The "Self-Repair" Method Another strategy involves activating stem cells already within the brain so that they migrate to damaged brain areas and replace the missing cells. This method, referred to as self-repair, gets around the difficulties associated with implantation while taking advantage of those stem cells already residing in the brain. But in order to make self-repair a viable option, neuroscientists must learn a great deal more about the normal process of neurogenesis. What is the chemical impetus that controls the "fate decisions" whereby a stem cell becomes one specific type of neuron rather than another? An answer to this question can only come from additional studies of neurons in culture media.

Dr. Jeffrey Macklis, of Children's Hospital in Boston, has already achieved a practical application of the neural stem cell research. He's discovered that he can induce the birth of new neurons in the cerebral cortex of mice from precursor cells already existing in the brain.

"And this was done without transplantation. The precursors were located near the fluid-filled ventricles. After damage to cells in the cortex, these precursor cells moved from the ventricular area to the area of brain damage. Once in place, the precursor cells progressively matured into fully developed neurons that remained for the remainder of the life of the animal. Not only that, but these new brain cells sent processes to connect to the original targets of the cells they were replacing. In short, we were able to induce in mice new nerve cells in the adult cerebral cortex—a site where new cells aren't customarily manufactured."

If Jeffrey Macklis's work in mice can be replicated in humans, treatments might become available for a host of patients with degenerative brain diseases. One of Macklis's own patients, Sally Carlson, developed Parkinson's disease while in her late 40's. The disease results from loss of dopamine-producing cells from a small nucleus in the brain stem called the substantia nigra, meaning "black substance," and, like most Parkinson's patients, Sally began having problems with movement. She found herself lacking energy; activities requiring mobility began to take longer. Other hallmarks of the disease—not incapacitating Sally so far—include tremors; a generally stooped, slowed posture; and increasing difficulty with managing buttons.

Although Parkinson's is a progressive disease, the rate of progression can be greatly slowed by drugs that are either converted within the brain into dopamine or that stimulate dopamine receptors in areas of the brain responsible for coordinating movement. Unfortunately, however, the drugs often lose their effectiveness after a few years and symptoms return and even progress. One solution, already mentioned, is to implant dopamine-producing cells within the brain. So far, clinical trials involving implantation have been beset by the twin difficulties of possible medical ethics violations (the stem cells may have been taken from the brains of aborted fetuses) and unacceptable side effects (many of the patients who received the implants developed unacceptable disabling side effects). Macklis's experiments with mice provide a less controversial and natural approach in stimulating some of the brain's own precursor stem cells.

Inducing Neurogenesis   According to Macklis, "If we can induce neurogenesis of one kind of neuron, that achievement should be generalizable to other kinds of neurons, such as the substantia nigra. If we're successful at generating new functional cells in the substantia nigra, we may be able to turn around, arrest, or slow down the progressive loss of dopamine-producing neurons in that area. While I don't think such a treatment is just around the corner, I believe over the next decade or so we're going to make substantial progress toward human therapies."

While obviously holding great promise for brain repair in Parkinson's disease, do neurogenesis and stem cell research represent the best hope for the other scourge among the brain's degenerative disorders, Alzheimer's disease?

"My hunch is that producing new nerve cells isn't the most important way of maintaining the brain at its best," says Marilyn Albert. "But the discovery of stem cells emphasizes that the brain is capable of growth and change even into late adulthood. A tiny but important part of that change might be the production of new nerve cells. But even more important is the brain's innate capacity for change in all of us no matter how old we are."

Neurogenesis will probably only be applicable to certain regions of the brain and will also be limited in terms of the number of new cells that can be produced. In addition, neurogenesis may only provide benefits in illnesses marked by comparatively simple circumscribed defects. In Parkinson's disease, for example, a deficiency of a single neurotransmitter, dopamine, results from atrophy of cells in that compact cluster of cells, the substantia nigra. Stem cell therapy might prove successful when directed at replacing these cells. Alzheimer's disease, in contrast, involves problems with many neurotransmitters and many defective cells that

Use It or Lose It

Like the infant brain responding to the newness of the
world, older brains respond to novelty by growing
new connections as well. Indeed, one of the keys to
retaining mental acuity in old age seems to be keeping a lively interest in a variety of activities—and especially learning new ones. Discovering an artistic bent,
going back to school, or taking up bicycling are all
good means to the same end: nourishing the neurons.

are spread throughout the brain rather than limited to a particular area. Thus,
Alzheimer's disease doesn't provide stem cells with a sufficiently defined target.

In addition, the early attempts at stem cell implantation have run into some
serious problems, as reported in 2001. Younger patients with Parkinson's disease
improved a bit after fetal stem cell implants, but only for about a year. After that,
they developed excessive and abnormal movements similar to those observed in
Parkinson's patients who take too many medications—an indication that the
implanted cells were producing too much dopamine. Even more troubling, the
abnormal movements persisted after the patients discontinued all drugs for
Parkinson's disease.

Fred Gage readily admits that stem cell research is only one avenue of
approach to the treatment of brain disease. But he also points out that exercise
and enriched environments appear to provide their benefits at least partially by
an increase in neurogenesis. "Both exercise and enriched environments lead to
a greater number of neurons surviving and maturing into functioning neurons.
In addition, the animals showed significant improvements in their behavior as
a result of the environmental stimulation."

Surviving and maturing is, of course, what we all would like to guarantee for our brains and ourselves as we head into the so-called twilight years. But even if we do all the right things and can count ourselves among the lucky ones who don't come down with some deadly disease, just how long can our brains be expected to survive? Although no one knows the answer to that question, I asked Fred Gage to speculate and give his best guess about brain longevity:

"I think it remains unknown whether or not there is an upper limit to brain survival. Most people die, not because their brain has aged, but because of deterioration somewhere else in the body. If we can maintain a healthy, functioning brain with continued self-renewal, I believe we can have a fully functioning or at least well-functioning brain for a longer period of time than our current estimates. It's important to remember that we have just begun to understand the functioning of the human brain."

EPILOGUE

Throughout our research and interviews for *The Secret Life of the Brain,* each of the participants told us of his or her sense of excitement and confidence about the future of brain research. And there is good reason for that excitement and confidence. Thanks to new and emerging technological advances, we can now study the brain at every level from human behavior to the swirl of molecules making up the components of neurotransmitters and their receptors. Thanks to the achievement of the first comprehensive analysis and interpretation of the human genome, announced in February 2001, the possibility—indeed the probability—now exists for discovering such secrets of the brain as the genes associated with manic depression, schizophrenia, and Alzheimer's disease. But new genetic insights are also likely to shed light on the functioning of the normal brain.

For instance, how does the brain of a genius differ from his or her less intellectually endowed counterparts? What is the neurobiological basis for temperament—that elusive personality trait that separates the introvert from the extrovert? Is creativity describable in terms of brain functioning? Many of these questions will, no doubt, be further understood thanks to brain research. Nevertheless, a few caveats seem in order.

Brain research differs from all other scientific enterprises in that the observer and the observed are one and the same. Brains are used to understand brains. While this doesn't necessarily imply a crucial impediment to our understanding, it does suggest the need for some humility about the ultimate limits of brain research.

Overreliance on one avenue of research to the exclusion of other avenues, for example, is unlikely to provide the comprehensive understanding that we seek. Physiologist William Rushton slyly captured the essence of the problem: Trying to discover how the brain works by poking electrodes into it (or by relying exclusively on any other single exploratory method) is like trying to discover the intentions of a foreign country by crossing its border on a dark night, entering a nearby town, and asking random passersby for their opinion.

What's needed to reveal more of the secrets of the brain is an all-encompassing theory of how the organ works. As we have learned from this book and the television series on which it is based, emphasis has largely shifted from structure to form. And why should that be surprising? Since the brain is an active, dynamic, supremely plastic structure that changes from moment to moment, it's likely that our understanding will also involve dynamic interpretations such as the one suggested in a marvelous analogy by the neurophysiologist Charles Sherrington:

"It is as if the Milky Way entered upon some cosmic dance. Swiftly the brain becomes an enchanted loom where millions of flashing shuttles weave a dissolving pattern, always a meaningful pattern though never an abiding one; a shifting harmony of subpatterns."

We have already discovered many secrets of this "enchanted loom." But we have surely only begun our search. The brain retains many more tantalizing secrets that within our foreseeable future will continue to challenge, puzzle, and fascinate.

INDEX

Down's syndrome, 166
Duffy, Frank, 34
Dyslexia, 56, 58-62, 64-65

E

Ecstasy (MDMA), 86-87
Ectoderm, 1, 2
Eden, Guinevere, 59, 60, 62, 64-65, 66, 67, 68
Electroconvulsive treatment (ECT), 130
Electroencephalograms (EEG), xx
Emotions. *See also specific emotions*
    adrenaline and, 125, 129-131
    and behavior, 27, 32, 73, 74, 75, 94
    cognition and, 73, 74, 75, 94, 109, 128, 129, 132,
        143-145, 150
    color-wheel analogy, 112
    deficiency of, 145-146, 148-150
    development in children, 24, 32, 35, 121-123
    feelings distinguished from, 146-150
    range of, 110-111
    and longevity, 132
    memory and, vii, 112, 113, 115, 125, 127, 130, 131,
        132, 141, 142, 156
    neural circuitry, 6, 7, 44, 111, 112-113, 116-117, 132,
        133, 142, 148, 150
    protein inhibitors and, 130-131
    stress and, 116, 118-131, 142
Empathy, 146-147, 148, 149
Endoderm, 1, 2
Entorhinal cortex, 163
Environment. *See* Gene–environment interactions
Epilepsy, 16
Exercise, 179-180

F

Faust, James, 182-183, 185
Fear response, 111, 114-130, 138
Feelings, emotions distinguished from, 146-150
Fetal alcohol syndrome, 16
Fetal brain development, 1-8, 11-18, 19, 29, 103
"Fight or flight" response, 115, 116, 121, 125, 138
Flashbacks, 119, 127
Focal magnetic stimulator, 183
Forebrain, 2, 3, 4
Fornix, 7, 113
Fovea, 21
Foveal cells, 21
Free radicals, 177-178
Freeman, John, 52, 53, 55
Friendship, 132
Frontal lobes, xv-xvi, 6, 32, 34, 42, 58, 73-74, 75, 76-77,
    79, 81, 83, 85, 89, 92, 98, 101, 125, 129, 132,
    138, 141, 144, 146, 156, 157, 160-162, 170

Functional magnetic resonance imaging (fMRI), xix, xx,
    49, 54, 60-62, 64-65, 67, 72, 87, 125, 126, 132,
    133, 161

G

Gage, Fred, 184-187, 190-191
Gage, Phineas, 146-147
Ganglion cells, 18
Gene–environment interactions
    in Alzheimer's disease, 166
    in brain development, 12-14-17, 26-34, 155
    in emotional states, 118, 135
    in dyslexia, 56
    in PTSD, 118, 121-122, 126-127
    in schizophrenia, 95-96, 97, 100
Giedd, Jay, 72, 73
Glia, 11-12, 175
Glutamate, 100-101
Gray matter, 5, 8, 33, 72, 100-101

H

Hallucinations, 96, 101
Happiness, 111, 131-133, 134
Hatten, Mary Beth, 4, 11-12, 14
Hearing, 6, 7, 19, 27, 29, 38-40
Hemispherectomies, 52-55, 57
Hindbrain, 2, 3
Hippocampus, 7, 84, 106, 113, 115, 132, 133, 137, 142,
    156, 157, 163, 180, 186
Hubel, David, 22, 24, 46
Humoral theory, 136-137
Hyman, Steve, 84, 85, 93-94
Hypothalamic-pituitary-adrenal axis, 121
Hypothalamus, 2, 7, 84, 113, 115, 121, 124-125, 129, 137

I

Ibuprofen, 175, 177
Infancy, ix, 1-34, 37, 38, 39-46, 44-45, 50, 74, 167
Inflammation, 175
Information processing, 160-162
Insula, 133, 150
Intelligence/I.Q., 8, 56, 58, 163
Isoniazid, 138-139

J

James, William, 28, 109
Johnson, Karen, 169, 170, 171
Justice, Ellen, 184

K

Kegl, Judy, 52-53
Kinoshita, June, vii
Kuhl, Pat, viii, 38-41, 45, 47
Kunitz, Stanley, x, 154-155, 165

CREDITS

Key: page numbers in boldface; (t) = top, (b) = bottom, (c) = center, (l) = left, (r) = right

Cover and front matter Mona Lisa, Bridgeman Art Library International Ltd.; **i** neural circuit, 422 Ltd.; **ii** neuronal connections, 422 Ltd.; **iv** cross section of the human cerebellum showing the Purkinje cells, Camillo Golgi (1843-1926) *Sulla fina anatomia degli organi centrali del sistema nervoso* (On the detailed anatomy of the central organs of the nervous system), Milan, 1886. Image courtesy of the Institute of the History of Medicine, Johns Hopkins University; **v** all images from PhotoDisc; **vii** filming The Secret Life of the Brain, Randi Anglin; **xvii** anatomical drawings of the brain, Johannes Vesling (1598-1649) *Syntagma Anatomicum* (The System of Anatomy), Padua, 1647. Image courtesy of the Institute of the History of Medicine, Johns Hopkins University; **xviii** (l) x ray, E. Decan, trauma.org, 6: *www.trauma.org/imagebank/imagebank.html;* (c) CT scan, Keith A. Johnson, M.D., Radiology and Neurology, Brigham and Women's Hospital, Harvard Medical School; (r) MRI scan, PhotoDisc; **xix** (l) PET scan, PhotoDisc; (r) fMRI scan, Diana Woodruff-Pak, Department of Diagnostic Imaging, Temple University, and Susan Lemieux, Department of Radiology, West Virginia University; **xx** (l) EEG, Patricia K. Kuhl, Ph.D., Speech and Hearing Sciences, University of Washington; (c) and (r) MEG data, CTF Systems Inc., a subsidiary of VSM MedTech Ltd., Canada.

Wider than the Sky xxii babies, Barbara Campbell/ Getty Images; **1** baby, PhotoDisc; **2-3** stages of fetal brain development, Kathryn Born; **4** scanning electron micrograph of spinal cord cells, courtesy of Kathryn Tosney, Ph.D., University of Michigan; **5** scans of cortex, Malcolm Godwin, Godwin@moonrunner.co.uk; **6-7** illustrations of brain anatomy, Kathryn Born; **9** neuron and synapse, Robert Finkbeiner; **10** neuronal connections, 422 Ltd.; **12** (l) scanning electron micrograph of neural crest cells, courtesy of Kathryn Tosney, Ph.D., University of Michigan; (r) neuron migrating along glial cell, David Grubin Productions; **13** functions of the cortex, Kathryn Born; **15** (tl) and (tr) depiction of ferret brain, 422 Ltd.; (bl) and (br) neuronal activity in ferret cortex, courtesy of Mriganka Sur, Ph.D., Department of Brain and Cognitive Sciences, Massachusetts Institute of Technology; **17** spectrograph images of axons and dendrites, David Grubin Productions; **20** all images from PhotoDisc; **23** babies with cataracts, David Grubin Productions; **28** premature baby in intensive care unit, PhotoDisc; **29** premature baby in special ICU, courtesy of Samantha Butler, Ph.D., Center for Neurobehavioral Studies, Children's Hospital, Boston; **31** baby, PhotoDisc; **32** baby expressions, all images from PhotoDisc.

Syllable from Sound 36 children in a row, EyeWire; **37** child on swing, PhotoDisc; **39** (l) baby in EEG cap and (r) EEG reading, David Grubin Productions; **40** neurons, from *The Postnatal Development of the Human Cerebral Cortex,* Vol I-VIII by J.L. Conel, Cambridge, Harvard University Press, 1939; **42** mother reading to child, Stephen Mautner; **43** language areas in the brain, Kathryn Born; **44** fMRI data of language activity in brain, courtesy of Marcus E. Raichle, M.D., Department of Radiology and Neurology, Washington University School of Medicine, St. Louis; **48** left and right hemispheres, Kathryn Born; **50** research from Laura Ann Petitto, Ph.D., Department of Education and Department of Psychology and Brain Science, Dartmouth College; photos by Jeffrey deBelle, *www.debellephotography.com;* **53** girl communicating in sign language, courtesy of Valerie Sutton, SignWriting Literacy Project, James Shepard-Kegl Nicaraguan Sign Language Projects, Inc.; **57** (t) Michael Rehbein, David Grubin Productions; (c) Michael with teacher, Randi Anglin; Michael's brain scans, courtesy of Dana Boatman, Ph.D., Department of Neurology, Johns Hopkins University; **58-59** brain regions associated with reading, 422 Ltd.; **63** child writing, PhotoDisc; **67** fMRI data of hyperlexic brain activity, courtesy of Guinevere F. Eden, Ph.D., Department of Neuroscience, Georgetown University.

A World of Their Own 70 adolescent boy, Peter Menzel/Stock, Boston Inc./PictureQuest; **71** teenage boy and girl, PhotoDisc; **73** teens on bus, PhotoDisc; **74** teen boy and girl, Elizabeth Kupersmith; **75** (t) data showing neuronal growth and pruning, courtesy of Paul Thompson, Ph.D., Laboratory of Neuro Imaging, University of California, Los Angeles; (b) illustration highlighting frontal lobe and amygdala, Kathryn Born; **76** teens playing volleyball, David Young-Wolff/Photo Edit/PictureQuest; **78** teenage boy, PhotoDisc; **80** PET scans of Ritalin, courtesy of *Journal of Neuroscience* 2001, Nora D. Volkow, M.D., Ph.D., Department of Medicine, Brookhaven National Laboratory; **84** reward pathway, Kathryn Born; **86** teens dancing, Michael Penland; **87** fMRI scans of brain on Ecstasy, John Nagy, National Institute of Drug Abuse/National Institutes of Health; **89** (l) fMRI data courtesy of Anna Rose Childress, Ph.D., Treatment and Research Clinic, University of Pennsylvania School of Medicine; (r) recovering cocaine addict, David Grubin Productions; **92** CT scans comparing brains of alcoholic and non-alcoholic, courtesy of Daniel Hommer, M.D., National Institute of Alcohol Abuse and Alcoholism; **95** Sabrina Yeskel, courtesy of the Yeskel family; **97** Alfred, Lord Tennyson, Library of Congress; **98-99** sequence on neuronal activity in hallucinations, 422 Ltd.; **100** CT scans comparing ventricle size, courtesy of Daniel Weinberger, M.D., National Institute of Mental Health, Clinical Brain Disorders Branch; **103** Isaac Walton, courtesy of the Walton family; **104** (t) and (b) Courtney Cook, Nancy Hale.

To Think by Feeling 108 visitor at the Vietnam Memorial, Stone 2000; 109 woman smiling, PhotoDisc; 111 data showing brain regions associated with emotions, courtesy of Antonio Damasio, M.D., Ph.D., Department of Neurology, University of Iowa College of Medicine; 112 (t) women hugging, PhotoDisc; (b) emotions color wheel, Annette DeFerrari; 113 brain regions that mediate emotions, Kathryn Born; 116 paths to amygdala, Leigh Coriale; 117 motorcycle race, Jean Pierre Boulme/Allsport/Vandystadt; 121 stress response, Kathryn Born; 125 Johnny Cortez, David Grubin Productions; 132 couple, PhotoDisc; 133 fMRI data on brain regions associated with love, courtesy of Andreas Bartels, Ph.D., Wellcome Department of Cognitive Neurology, University College London; 135 Lauren Slater, Amanda Pollak; 136 depression, PhotoDisc; 138 neurotransmitter sources, Leigh Coriale; 143 serotonin reuptake, Robert Finkbeiner; 144 PET scans showing bipolar disorder, courtesy of John Mazziotta, Ph.D., Laboratory of Neuro Imaging, University of California, Los Angeles; 145 (t) Robert Schumann, Library of Congress; 147 Phineas Gage images, courtesy of Warren Anatomical Museum, Francis A. Countway Library of Medicine, Harvard Medical School; 148 brain before and after stroke, 422 Ltd.; 149 Marvin and Arlene Bateman, David Grubin Productions.

Second Flowering 152 Mstislav Rostropovic, David Burnett/Contact Press Images/PictureQuest; 153 older couple, PhotoDisc; 155 (l) Stanley Kunitz at 95, Ted Rosenberg, W.W. Norton & Company; (r) Stanley Kunitz in 1962, LaVerne Harrell Clark; 156 (t) older couple, Elizabeth Kupersmith; (b) types of memory, Leigh Coriale; 157 brain regions associated with memory, Kathryn Born; 159 neurons in normal aging vs. Alzheimer's disease, Leigh Coriale; 161 fMRI data comparing older and younger people doing different tasks, courtesy of Denise C. Park, Ph.D., Center for Applied Cognitive Research on Aging, University of Michigan; 162 tennis players shaking hands, PhotoDisc; 164 slides comparing NMDA receptors in older and younger monkeys, courtesy of John Morrison, Ph.D., Neurobiology of Aging Laboratory, Mount Sinai School of Medicine, New York; 166 baby and grandmother, PhotoDisc; 169 PET scans of activity in normal brain vs. Alzheimer's disease, Laboratory of Neurosciences, National Institute on Aging; 172 plaques and tangles, David Grubin Productions; 173 illustrations of beta-amyloid plaques, 422 Ltd.; 177 tau tangles, 422 Ltd.; 178 Red Simmons, Edward Gray; 179 Milton Adamson, Edward Gray; 184 CI movement therapy, Edward Taub, Ph.D., University of Alabama, Birmingham; 185 James Faust, Edward Maritz; 186 neurons, David Grubin Productions; 187 neurogenesis, courtesy of Fred Gage, Ph.D., Laboratory of Genetics, Salk Institute for Biological Studies; 190 all images from PhotoDisc.

Production credits for the PBS documentary THE SECRET LIFE OF THE BRAIN

Executive Producer
David Grubin
Producers
David Grubin
Edward Gray
Tom Jennings
Michael Penland
Amanda Pollak
Co-Producers
Sarah Colt
Annie Wong
Associate Producer
Jenny Carchman
Editors
Seth Bomse
Nobuko Oganesoff
Deborah Peretz
Josh Waletzky
Cinematography
James Callanan

Gregory Andracke
Edward Marritz
Music
Michael Bacon
Narrator
Blair Brown
Production Executive
Lesley Norman
Animation
422 Ltd.
Sound Recording
Roger Phenix
David Gladstone
Matthew Magratten
Mark Mandler
Assistant Editors
Amy Ulrich
Rachel Hepler
Office Manager
Sara Levine

Series Production Coordinator
Shira Loewenberg-Dokic
Series Co-Producer
Sarah Colt
Science Editor
June Kinoshita
Advisors
Jenny Atkinson
Carl W. Cotman, Ph.D.
Patricia S. Goldman-Rakic, Ph.D.
Steven E. Hyman, M.D.
Story C. Landis, Ph.D.
Bruce S. McEwen, Ph.D.
J. Anthony Movshon, Ph.D.
Steven Petersen, Ph.D.
Dorothy Suecoff
Michael Templeton

Program Development
David Grubin
June Kinoshita

For WNET
Project Management
Merle Kailas
Coordinating Producer
Jared Lipworth
Executive in Charge
William R. Grant
Executive Producer
Beth Hoppe

THE SECRET LIFE OF THE BRAIN

A co-publication of The Dana Press and the Joseph Henry Press.

Printed in the United States of America

The Dana Press, a division of The Charles A. Dana Foundation, publishes health and popular science books about the brain for the general reader. The Dana Foundation is a private philanthropic organization with particular interests in health and education.

The Joseph Henry Press, an imprint of the National Academy Press, was created with the goal of making books on science, technology, and health more widely available to professionals and the public. Joseph Henry was one of the founders of the National Academy of Sciences and a leader in early American science.

Quotations from the poems of Stanley Kunitz appearing on pages x and 154 of this volume are reprinted courtesy of W.W. Norton and Company, New York, NY.

Library of Congress Cataloging-in-Publication Data

Restak, Richard M., 1942-
    The secret life of the brain / Richard Restak.
        p. cm.
    Includes index.
    ISBN 0-309-07435-5 (alk. paper)
    1. Brain—Popular works. 2. Neurosciences—Popular works. I. Title.

QP376 .R4729 2001
612.8'2—dc21

2001039777

Major funding for the PBS documentary The Secret Life of the Brain is provided by the National Science Foundation. Corporate funding is provided by Pfizer Inc and The Medtronic Foundation on behalf of Medtronic, Inc. Funding is also provided by the Park Foundation, PBS, the Corporation for Public Broadcasting, The Dana Foundation, and the Dana Alliance for Brain Initiatives.

Managing editors	Stephen Mautner
	Jane Nevins
Project editor	Roberta Conlan
Designer	Francesca Moghari
Assistant editor	J. Andrew Cocke
Production manager	Dorothy Lewis
Illustrations research	Susan S. Blair
	Elizabeth Cook Thompson
Art coordinator	Randy Talley
Printer	Chroma Graphics, Inc.
	Largo, Maryland

Scientific consultants   Robert Cook-Deegan, M.D.
Institute of Medicine
Washington, D.C.

Carl Cotman, Director
Institute for Brain Aging and Dementia
University of California, Irvine
Irvine, California

Murray Goldstein, Medical Director
United Cerebral Palsy Research
 and Educational Foundation
Washington, D.C.

Barry Gordon, M.D., Director
Department of Neurology and
 Cognitive Science
Johns Hopkins Medical Institutions
Baltimore, Maryland

Gary Gottlieb, M.D.
Professor of Psychiatry
Harvard Medical School
Boston, Massachusetts

Annette Karmiloff-Smith, Head
Neurocognitive Development Unit
Institute of Child Health
London, England

Karin Nelson, M.D., Acting Chief
Neuroepidemiology Branch
National Institute of Neurological
 Disorders and Stroke
Bethesda, Maryland

Judith Rapoport, M.D., Chief
Child Psychiatry Branch
National Institute of Mental Health
Bethesda, Maryland